D0899166

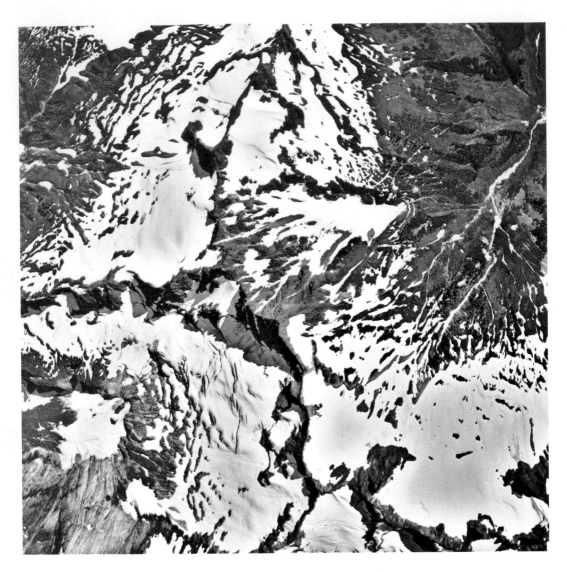

Mount Buckindy in the Cascades of northern Washington is a coincident color and copper soil anomaly *(U.S. Forest Service photograph).*

Geology of the
Porphyry Copper Deposits
of the Western Hemisphere

By

Victor F. Hollister

Published by

Society of Mining Engineers

of

The American Institute of Mining, Metallurgical, and Petroleum Engineers, Inc.

New York, New York • 1978

Preface

Porphyry copper deposits of the Western Hemisphere occur in clusters whose characteristics reflect their crustal and tectonic setting. Clusters commonly occur in elongate belts near plate boundaries, but subduction has a variable genetic influence in the six different porphyry copper provinces of the western hemisphere.

Evidence cited in this volume suggests that porphyry copper development is not necessarily restricted to subduction zones. The timing for such a genetic tie is frequently poor; the spatial relations are not everywhere convincing; and the lead isotopic data are not compatible with derivation of the pluton from a subducted oceanic plate.

A zone of partial melting tapped by a deeply penetrating fault is, however, proposed as a favorable germination point for a porphyry copper. If the fault is the Benioff zone, and the zone of partial melting is on or adjacent to it, then a porphyry deposit conceivably may be generated in a subduction zone. A deeply penetrating fault of any type may be adequate to generate a porphyry copper deposit, however, providing the fault penetrates a zone of partial melting.

In its surface expression, however, a porphyry system may occur either as a piercement or as a complex controlled by a major fault. Piercements that do not lie on major structures indicate that the porphyry magma and fluids may penetrate the crust at will, once they are formed.

Deposits are described in terms of suggested modifications to the Lowell and Guilbert (1970) model. As presented in this volume, the Lowell and Guilbert model describes deposits in which the quartz-bearing mineralized pluton usually is associated with a concentrically zoned potassic-phyllic-argillic-propylitic alteration sequence. Mineralogy in each assemblage varies from one belt to another, and one or more alteration zones may be missing in specific deposits.

Deposit features in the Andean orogen are predictable by the Lowell and Guilbert (1970) model. Deposits of the Appalachian orogen fit only if the model is modified to reflect greater erosion in the region. Deposits south of the Columbia plateau in the Cordilleran orogen commonly are consistent with this model where a Precambrian basement is inferable.

Deposits associated with quartz-bearing igneous host rocks in the Caribbean, in much of Alaska, and in most of the Cordilleran orogen north of the Columbia plateau, are not consistent with the Lowell and Guilbert (1970) model unless the igneous and alteration mineral assemblages are substantially modified. Potassic, phyllic, argillic, and propylitic alteration zones generally characterize this model, although exceptions (e.g., Ajo) are known.

Porphyry copper deposits in quartz-free igneous rocks commonly contain only potassic and propylitic alteration zones. The igneous host rocks may be alkalic or calc-alkalic, and diorite commonly is prominent in the intrusive center. The diorite model, proposed to define porphyry copper deposits that occur with quartz-free hosts, has been found to be economically useful in some terranes. Most diorite-type porphyry copper deposits are of the copper-gold type, whereas deposits of the Lowell and Guilbert (1970) model may be either copper-gold or copper-molybdenum type.

Porphyry copper deposits with a high molybdenum:copper ratio ordinarily are found where the mineralized pluton penetrates either a cratonic crust or a crust with a thick sialic component, as in the Andean, Appalachian, or southern Cordilleran orogen. Deposits with high gold:copper ratios commonly occur in areas with thick sections of basic volcanic rocks, e.g., the Caribbean, the Cascades, or the continental margin of Alaska. However, exceptions to these generalizations are well known (e.g., Pyramid in the Aleutian arc has a normal molybdenum: copper ratio, and Bisbee has a high gold: copper ratio). It is inferred, therefore, that some molybdenum and gold must derive from the crust. As copper appears in both models, the source of this metal may be largely subcrustal.

Copper, gold, and molybdenum may occur together in a potassic zone (as at Bingham); or any or all of these elements may be transported out of the potassic zone by highly saline magmatic-hydrothermal fluids, developing gold-rich phyllic or argillic zones (e.g., Pueblo Viejo and Battle Mountain), molybdenum-rich phyllic zones (e.g., Cat-heart), and copper-barren potassic zones (e.g., Pyramid).

Tectonic style also influences the type of porphyry copper deposit formed. Alkalic mineralized intrusions have been found only in the Triassic-Lower Jurassic rifted terrane of the northern Cordillera. Quartz-bearing mineralized plutons most commonly occur associated with strike-slip faults, and this type of deposit is dominant in most porphyry copper provinces.

Tourmaline and breccia columns form a common association in Lowell and Guilbert (1970) model deposits, and they occur in all regions with porphyry copper occurrences.

Table of Contents

Frontispiece: Mount Buckindy in the Cascades of Northern Washington

Figures

Tables

Plates

Introduction

This volume summarizes characteristics and the geological setting of porphyry copper deposits of the Western Hemisphere. Grouped by characteristics, porphyry copper deposits are described in six regions: Andean orogen, Appalachian orogen, Alaska, northern Cordilleran orogen, southern Cordilleran orogen, and the Caribbean. Each region has its own peculiarities of geologic setting, and those deposit characteristics that may be projected from one chapter to another will be used in reaching the conclusions from which to draw genetic models for porphyry copper deposits.

The term porphyry in this volume denotes a type of ore deposit, rather than a rock classification. The classification of intrusive igneous rocks used in this volume is shown in Figs. 36a and 42 (pp. 103, 130). Some modern classifications suggest mineral-ratio definitions other than those used in this volume; however, most literature references cited are based on the classification shown in Figs. 36a and 42. The reader should be aware, therefore, that quartz monzonite in this volume contains from 35% to 65% potassuim feldspar and at least 10% quartz.

A term used in this volume is defined where introduced within the context of a specific chapter. Distensional tectonic conditions are used in the sense of Cady (1972) in the Appalachian orogen, for example. Porphyry copper deposits are defined as large bodies of mineralized rock, usually including a porphyritic intrusive phase, that have copper sulfide minerals disseminated through them. The definition has no commercial connotation and is used strictly in a noneconomic sense to encompass the spectrum of copper mineralization associated with these deposits. Terms commonly accepted in the literature are not defined.

Porphyry molybdenum and porphyry copper deposits commonly are distinguished by most economic geologists. The basis for the separation is summarized in one chapter of this volume.

To permit the reader to more easily peruse the bibliography for any particular region, references will be found at the end of each chapter. The bibliography is not intended to be complete, but is designed to provide adequate minimum coverage for descriptions of deposits in the chapter concerned. Each entry is properly referenced in the text of that chapter to facilitate the reader's research into any particular topic.

This volume is written from the point of view of the exploration geologist whose task it is to evaluate the structural, petrographic, mineralogic, and metallogenic characteristics of a porphyry copper deposit. The goal of the evaluation is to determine if, and where, concentrations of metals may exist in an igneous-hydrothermal system that includes the most complex set of geologic factors found in any of the many types of ore deposits. The successful porphyry copper geologist is simultaneously a petrographer, geochemist, structural geologist, and mineralogist.

Some porphyry copper deposits have developed in volcanic arcs underlain by a Benioff seismic zone with a trench paralleling the arc. The most convincing example of a deposit known to have developed in an environment where subduction is an active process is the 3.3 m.y. (million years) old Dry Creek occurrence in the Aleutian arc (this volume). On the other hand, the younger deposits in the Cascades appear to have formed in a volcanic arc lacking both the Benioff zone and the offshore trench. Subduction is not widely accepted as an active mechanism for the period when the younger Cascade deposits formed. Subduction has also been questioned for deposits formed at a great distance from a plate boundary (e.g., Bingham). Regardless of the pros and cons regarding subduction, most deposits are associated with regional faults. Strike-slip faulting most commonly is identified with deposits where a major fault is present. The preference of porphyry copper deposits for orogenic belts near plate boundaries implies a genetic tie between failure near a plate margin by strike-slip faulting and porphyry copper formation. The failure may occur whether or not subduction is an active process. Support for subduction varies in different regions and therefore this volume does not emphasize subduction's role in porphyry copper deposition.

Porphyry copper deposits have been classified as copper-molybdenum or copper-gold types on the basis of the coproducts present. Brenda is a deposit with a high molybdenum: copper ratio. On the other hand, Pueblo Viejo, Bisbee, and Battle Mountain have abnormally high gold:copper but very low molybdenum:copper ratios. In general, porphyry copper deposits that produce byproduct molybdenum, or have molybdenum:copper ratios above 0.02, seldom have a gold: copper ratio in excess of 0.57 g (0.007 oz) Au to the percent copper. On the other hand, deposits with gold:copper ratios in excess of 0.28 g (0.01 oz) Au to the percent copper commonly have molybdenum:copper ratios under 0.005. This classification fails, however, when deposits with both high gold:copper and high molybdenum:copper ratios are considered (as with Bingham).

Porphyry copper deposits could be classified by the dominant structure within the deposit, since some (e.g., Toquepala and Braden) are essentially mineralized breccia columns, while others (e.g., Chuquicamata and Esperanza) are stockworks. Confusion may occur with such inflexible modeling, since minor breccia pipes are found in stockwork deposits, and stockworks can be found peripheral to major breccias, such as Toquepala and Braden.

The volume also could have been organized by metasomatic products found with hypogene copper sulfide. An orthoclase-bearing potassic zone classically is developed in most Andean deposits. On the other hand, a sodium-rich metasomatic zone appears in many rift-associated porphyries, in some continental-margin deposits, and in a few of those in the island-arc environment. However, some deposits have both secondary biotite (containing potassium) and albite, but no orthoclase in a potassic zone; and the existence of both secondary sodic as well as potassic minerals prohibits clear separation based on alkali metals present in secondary silicates.

The volume could be organized according to petrographic association, because both alkalic and calc-alkalic plutons are associated with porphyry copper deposits. CIM Special Vol. 15, a compilation of porphyry copper deposits of the Canadian Cordillera, distinguishes between porphyry copper deposits with alkalic and calc-alkalic associations. Elsewhere, for example in Alaska, alteration makes it impossible to determine if basic igneous host rocks are alkalic or calc-alkalic, so that an alkalic:calc-alkalic classification is not universally practical.

On the other hand, alteration mineral assemblages described for alkalic igneous rocks associated with porphyry copper deposits in the Canadian Cordillera are similar to alteration mineral assemblages developed in calc-alkalic, nonquartz-bearing hosts. Diorite (or its extrusive equivalent) is dominant in most of these deposits, whether alkalic or calc-alkalic, if the porphyry is associated with a nonquartz-bearing host. Alteration mineral assemblages for dioritic hosts rarely include the phyllic and argillic zones so common in porphyry copper deposits associated with quartz-bearing igneous rocks, as described by Lowell and Guilbert (1970).

Copper Mountain, Afton, and Galore Creek are cited as examples of an alkalic diorite model deposit; whereas Baultof, Alaska, is given as a calc-alkalic diorite occurrence in this volume. In both cases, copper occurs in potassic and propylitic alteration sequences. Phyllic zones are absent from these deposits.

The simplest mechanism for categorizing is a separation of porphyry copper deposits into two models: a diorite model to include the quartz-deficient type, which contains only a potassic and propylitic zone; and a modified Lowell and Guilbert model expanded to include copper-gold and copper-molybdenum porphyry deposits displaying the aberrations found when the host is a calc-alkalic pluton with a high $Na_2O:K_2O$ ratio, or is deeply eroded, or metamorphosed. The two models present distinct ore controls and reflect different crustal conditions.

Tables summarize the characteristics of porphyry copper deposits in each of the separate chapters covered in this volume. The tables permit quick cross reference of characteristics from one area to another. Not all tables contain the same headings, however, because the tables compiled for this volume are intended to summarize data available in the literature. For this reason, tables on deposits of the Alaskan, Andean, Appalachian, and the northern Cordilleran orogen offer some difference in data than tables for other areas. Deposits in these four areas are not described as commonly in the literature as are porphyries from the southern Cordilleran orogen.

ACKNOWLEDGMENTS

A. W. Rose, M. Wolfhard, and S. W. Hobbs were very helpful in providing stimulating discussions on the interpretation of field data, and their help should be recognized. W. R. Bergey provided critical comment on some portions of the text, and it is hoped that his views are adequately incorporated. R. B. Campbell and G. W. Walker provided many thoughtful comments that strengthened the volume. C. H. Burgess was encouraging in providing early impetus for the compilation of the material for this volume, and his help is gratefully acknowledged. E. M. Mac-Kevett, Jr. has also provided many helpful criticisms, which are gratefully accepted. L. H. Hart was especially helpful in the past for his descriptions of individual deposits. It is a pleasure to acknowledge his help. J. E. Armstrong, J. M. Carr, A. J. Sinclair, C. I. Godwin, D. W. Heddle, P. W. Guild, J. D. Lowell, and S. J. Haynes provided helpful criticism and their contributions have strengthened the volume.

J. R. Woodcock provided much original material for the chapter on molybdenum, and therefore he is coauthor of this section. Failing to acknowledge his help in this manner would inadequately reward his contribution.

Porphyry Copper Deposits of the Andean Orogen

CONTENTS

INTRODUCTION

The regional characteristics of porphyry copper deposits in South America southward from Pantanos and Pegadorcito, Columbia, will be summarized. The age of formation of deposits spans the period from the Permian to the present. Where possible, the general characteristics and the individual structural features of the deposits are integrated into the plate tectonic model. Individual deposits cited in this chapter, located in Fig. 1, are the best known and most fully described, and form the Andean copper belt (Peterson, 1958). Smaller porphyry copper deposits fall both east and west of the line of large, high grade deposits (e.g., Chuquicamata and Toquepala) that make up the spine of this belt. Most of those smaller deposits (e.g., Mi Vida [Koukharsky and Mirre, 1976]) are omitted from Fig. 1 and Tables 1 and 2, however. Neither the Andean orogen nor the individual deposits within it are as well described in the literature as are their counterparts in the Cordilleran orogen of North America. Therefore, generalizations relying on regional geologic mapping, geophysics, and isotopic geochemistry lack details pertinent to this type of deposit in the Cordilleran orogen.

The South American porphyry copper province has been described by Hollister (1974) to include most of the Andean orogen excepting the tin province of Bolivia. Porphyry copper deposits have not been commonly found within the tin belt, although copper-tin (e.g., San Rafael, Peru) and copper-tungsten (e.g., Llamuco, Chile) deposits are known. James (1971) shows the depth to the Moho to be appreciably greater where the tin belt exists, but disclaims any genetic significance from the coincident appearance of the tin-tungsten province and the inhibition of porphyry copper development where the sialic crust thickens. Dates on tin mineralization in this belt (Clark and Farrar, 1973) are similar to some dates for porphyry copper mineralization known elsewhere in the Andean orogen. It seems that some mechanism is needed to explain the geographic separation of porphyry copper-molybdenum and simultaneous tin-tungsten-molybdenum metallization within the same tectonic system and appearing at similar distances from the trench. The thickened crust of the tin belt is one of the few known differences.

Tables 1 and 2 list some of the most important porphyry copper deposits explored to date, but do not include Antaminas, Peru, or a number of other large skarn deposits

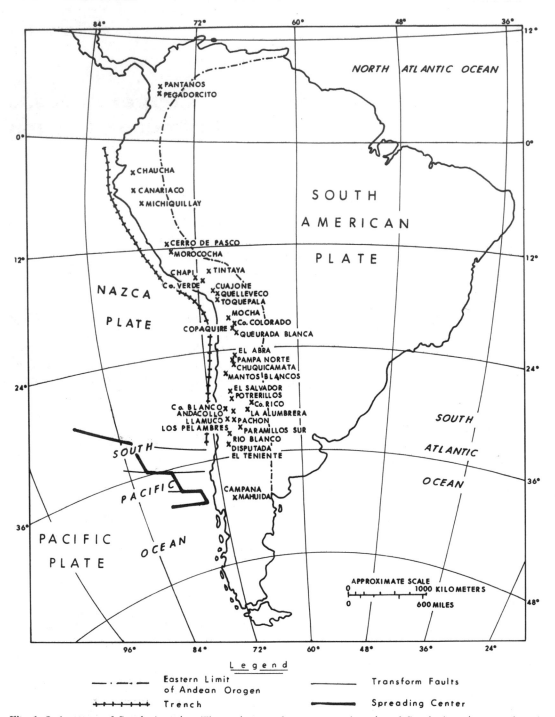

Fig. 1. Index map of South America. The major porphyry copper deposits of South America are plotted on the plate tectonic base of US Geological Survey map MF340. The approximate eastern limit of the Andean orogen is added. The richer and larger deposits tend to form a linear pattern that is subparallel to the trench.

associated with intrusions. In several of the latter, porphyry-type mineralization exists in the intrusion to some extent. Also not included are "special case" deposits such as Cerro de Pasco, where sulfide mineralization does occur in, and spatially related to, an intrusion. Suspected possible porphyry-type occurrences are omitted, as are many deposits with poorly known characteristics. Porphyry systems having two names for different parts of the same deposit are simplified (e.g., Santa Rosa is combined with Cerro Verde). Some names for newly discovered parts of previously known deposits are omitted if the older names are well known and used in the literature (e.g., Morococha). As a result of the parameters used in guiding the selection of deposits for incorporating in these tables, those listed are now among the most completely explored and adequately described mineralized intrusions in the Andean orogen.

GEOLOGIC SETTING

Regional Geology

The following summary makes mention of those aspects of the regional geology of the Andean orogen that seem to pertain particularly to the genesis of the porphyry copper deposits. For more complete descriptions of the geology, see James (1971), Ericksen (1975), or Ocola and Meyer (1973), or references cited by them. Fig. 2 gives the location of the plutonic and Precambrian metamorphic rocks and shows gravity data for the Andean copper belt. From these data, the reader may gain a reconnaissance impression of the crust.

From Campana Mahuida north at least to Chaucha, the porphyry copper province is characterized by mineralized plutons that intrude a crust which includes both old (pre-Mesozoic) and young (Mesozoic and Tertiary) rocks. The pre-Mesozoic rocks are composed largely of crystalline complexes in the vicinity of the porphyry deposits and the younger rocks are mostly arc-trench assemblages or continental volcanic and sedimentary rocks, although platform and shelf sediments are locally prominent in the post-Paleozoic environment of some deposits.

Pre-Mesozoic: The pre-Mesozoic rocks include both Precambrian and Paleozoic assemblages. Gneiss with Precambrian radiometric dates occurs in Arequipa near Cerro Verde, near Puira, northern Peru, and in various other locations fairly close to some of the larger known porphyry deposits. The reported Precambrian (Proterozoic) gneiss, schist, and plutonic rocks suggest that this part of the ancient crust was dominantly sialic in composition. Incomplete descriptions of the Precambrian imply a predominance of metasedimentray material in the nonplutonic rocks; the plutons are granodioritic. The presence of the Precambrian in the heart of the Andean porphyry copper belt suggests that many of the mineralized plutons passed through Precambrian crystalline rocks as they rose in the crust.

Paleozoic rocks, where present in the vicinity of porphyry copper deposits, are mainly nonplutonic and consist largely of metasediments with volcanic rocks only locally important. Widespread Devonian marine shelf sediments are found in Peru, Bolivia, and Chile unconformably overlying lower Paleozoic and Precambrian rocks. Carboniferous and Permian sedimentary and volcanic rocks are widely distributed within the porphyry copper province as are granodiorite plutons of the same age (Fig. 3).

Specific examples of the association of Precambrian and Paleozoic sedimentary, volcanic, and plutonic rocks are cited. Chaucha, Ecuador, lies on the boundary of a thick continental crust to the east (which probably includes Precambrian rocks) and a thick Mesozoic basic marine volcanic sequence in the upper crust to the west (Goosens and Hollister, 1973).

Michiquillay (Hollister and Sirvas, 1974) has nearby sparse outcrops of Permian, Carboniferous, and Precambrian schists erratically interspersed with younger Cretaceous marine sedimentary rocks.

The porphyry copper bearing plutons in the Quelleveco-Cuajone-Toquepala areas probably penetrated Precambrian crystalline rocks as well as Paleozoic, Mesozoic, and Tertiary rocks (James, 1971).

The region from the Copaquire deposit to Chuquicamata, Chile, also contains possible

Table 1. Stockwork-Type Porphyry Copper Deposits of South America

Deposits	K-Ar Age	Intrusive Close to Ore	Rock Intruded	Hypogene Grade, %	Primary Regional Fault	Secondary Regional Fault	Alteration Zones Present	Pyritic Halo Dimension, km	Zoned District	Remarks*
Argentina										
Campana Mahuida Neuguen	74.2	Qtz Mon Por	Cret Sed	0.8 Cu	N35E		Phy Arg Prop	3 x 2	Ag-Pb-Zn-Cu	Reserve 20 MT @ 0.8% Cu
La Alumbrera Catamarca	8.0	Qtz Mon Por	Tert Volc	0.4 Cu 0.04 Mo	N45W		Phy Arg Prop	2 x 1	None	Reserve 5 MT @ 0.5% Cu
Paramillos Sur. Mendoza	Trias (?)	Qtz Mon Por	Per Tria Sed & Vol	0.38 Cu 0.02 Mo	N-S		Potassic Phy Arg	3 x 2	Pb-Zn-Cu	Reserve 105 MT @ 0.38 Cu
Chile										
Chuquicamata, Anto	29.2	Qtz Mon Por	Grano-diorite	1.2 Cu 0.04 Mo	N10E		Pot Phy Arg Prop	8 x 4	Pb-Zn-Cu-Mo	Reserve 1400 MT @ 1.2 Cu
Mocha Tarapaca	56.4	Qtz Dio Por	Tert Volc	0.6 Cu 0.03 Mo	N40W		Phy Arg Prop	4 x 3 (?)	None (?)	Reserve 107
El Salvador Atacama	39.1	Grdr Por	Tert Volc	0.9 Cu 0.04 Mo	NA		Pot Phy Arg Prop	3 x 3	None (?)	Reserve 260
Potrerillos Atacama	34.1	Grdr Por	Jura Sed	1.6 Cu 0.03 Mo	NA		Pot Phy Arg Prop	6 x 4	Pb-Zn-Cu	Reserve 30 MT @ 1.65 Cu
Co Colorado Tarapaca	Tert	Dac Por	Jura Sed	NA	N-S		Pot Phy Prop	4 x 2	None (?)	Reserve 100 MT @ 1.2 Cu
Copaquire Anto	22 (?)	Qtz Mon Por	Jura Sed	NA	N-S	N20W	Phy Arg Prop	4 x 2	None	High Mo section
Andacollo Coquimbo	90 (?)	Grdr	Jura Volc	0.6 Cu 0.015 Mo	N20W	N40E	Pot Phy Arg Prop	5 x 2.5	Pb-Zn-Cu-Mo	Reserve 350 MT @ 0.7 Cu
Pampa Norte Anto	29 (?)	Qtz Mon Por	Grdr	NA	N10E		Pot Phy Arg Prop	5 NSx 2 EW	None	Reserve 260 MT @ 0.7 Cu
Qbda Blanca Anto	Tert	Qtz Mon Por	Meso Volc	NA	N-S		Pot Phy Arg Prop	6 x 4	Pb-Zn-Cu	Reserve 200 MT @ 1.0 Cu
Mantos Blancos Anto	Tert?	Dac Por	And Volc	0.7 Cu Tr Mo	NA		Phy Arg Prop	3 x 1	None	Reserve 10 MT @ 1.8 Cu in 1960
Columbia										
Pegadorcito	Tert	Qtz Dio Por	Tert Volc	0.72 Cu	NA		Phy Arg Prop	5 x 5	NA	Reserve 500 MT @ 0.7 Cu(?)
Pantanos	Tert	Qtz Dio Por	Tert Volc	0.76 Cu	NA		Phy Arg Prop	5 x 5	NA	
Ecuador										
Chaucha, Azuay	9.9	Qtz Mon Por	Grdr	0.7 Cu 0.03 Mo	E-W	N10E	Phy Arg Prop	4 x 3	None	Reserve 75 MT @ 0.7% Cu

Table 1. Stockwork-Type Porphyry Copper Deposits of South America—Continued

Deposits	K-Ar Age	Intrusive Close to Ore	Rock Intruded	Hypogene Grade, %	Primary Regional Fault	Secondary Regional Fault	Alteration Zones Present	Pyritic Halo Dimension, km	Zoned District	Remarks[*]
Peru										
Michiquillay	20.6	Qtz Mon Por	Cret	0.6 Cu 0.02 Mo	N45W	N50E	Phy Arg Prop	2 x 2	None	Reserve 575 MT @ 0.72 Cu
Tintaya (Quechua)	Tert	Qtz Mon Por	Qtz Tert Volc	NA	N50W	NA	Pot Phy Arg Prop		Pb-Zn-Cu	Reserve 100 MT @ 0.6 Cu
Quelleveco	Lower Tert	Qtz Mon Por	Cret Volc	0.6 Cu 0.03 Mo	E-W	NA	Phy Arg Prop	3 x 1.5	Pb-Zn-Cu	Reserve 200 MT @ 0.95% Cu
Canariaco, Piura	Tert	Qtz Mon Por	Cret Sed	NA	N-S	NA	Pot Phy Arg Prop	3.5 x 2	None (?)	Reserve 200 MT @ 0.5 Cu
Morococha (Toro Mocho)	7.0	Qtz Mon Por	Perm Cret	0.7 Cu 0.02 Mo	N25W	E-W	Phy Arg Prop	5 x 3	Ag-Pb-Zn Cu-Mo	Reserve 200 MT @ 0.76 Cu
Cuajone	Tert	Qtz Mon Por	Cret Volc	0.7 Cu 0.03 Mo	N60W	E-W	Pot Phy Arg Prop	5 x 4	Pb-Zn-Cu	Reserve 1300 MT @ 1.0 Cu

Jura—Jurassic
Cret—Cretaceous
Tert—Tertiary
Trias—Triassic
Qtz—Quartz

Mon—Monzonite
Por—Porphyry
Volc—Volcanic
Sed—Sediments
Dio—Diorite

Dac—Dacite
Phy—Phyllic
Pot—Potassic
Arg—Argillic
Prop—Propylitic

MT—Million Tons
NA—Not Available

[*]Tons (US short) × 0.91 = metric ton

Table 2. Tourmaline Breccia Pipe Deposits of South America

Deposits	K-Ar Age	Intrusion Close to Ore	Rock Intruded	Hypogene Grade, %	Primary Regional Fault	Alteration Zones Present	Pyritic Halo Dimension, km	Zoned District	Remarks*
Argentina									
Pachon, San Juan	Tert	Qtz Mon Por	Tert Volc	0.65 Cu / 0.015 Mo	None	Phy Arg Prop	8 x 6	None(?)	Reserve 550 MT @ 0.65 Cu
Chile									
Los Pelambres, Coquimbo	9.96	Grdr Por	Grdr sed & Volc	0.7 Cu / 0.03 Mo	None	Pot Phy Arg Prop	6.5 x 2.5	PbZnCu	Reserve 428 MT @ 0.78 Cu
Rio Blanco (Andina) Santiago	4.59	Qtz Mon Por	Tert Volc	1.34 Cu / 0.03 Mo	N15W	Phy Arg Prop	5EW x 11NS		Reserve 3000 MT @ 1.24 Cu tourmaline-rich ore
Disputada Santiago	same?	Dacite	Grdr Volc	1.7 Cu	None	Phy Arg Prop	same		Reserve 50 MT 1.7 Cu
El Teniente (Braden) O'Higgins	4.32	Dacite	Tert volc	1.7 Cu / 0.05 Mo	None	Pot Phy Arg Prop	3.5 x 4.0	Sb-Ag-Pb Zn-Cu	Reserve 4000 MT @ 1.05 Cu
El Abra, Anto	33.2	Qtz Mon Por	Paleo Sed	0.9 Cu / 0.03 Mo	N10E	Phy Arg Prop	3.5 x 3	None(?)	Reserve 1500 MT @ 1.09 Cu
Peru									
Toquepala	58.7	Dacite	Cret volc	0.7 Cu / 0.04 Mo	None	Phy Arg Prop	3 x 3	None	Reserve 400 MT @ 0.99 Cu in 1956
Chapi	Tert	Dacite	Grdr	1.0 Cu / 0.03 Mo	None	Pot Phy Arg Prop	1.5 x 1.5	None	Reserve 5 MT @ 2.0 Cu
Cerro Verde	58.8	Qtz Mon Por	Grdr	NA	None	Pot Phy Arg Prop	3 x 3	Ag-Pb-Zn-Cu	Reserve 1200 MT @ 0.67 Cu (include Sta. Rosa)

NA—Not Available
Grdr—Granodiorite
Tert—Tertiary

Qtz—Quartz
Mon—Monzonite
Por—Porphyry

Sed—Sediments
Paleo—Paleozoic
Volc—Volcanic

Pot—Potassic
Phy—Phyllic
Arg—Argillic
Prop—Propylitic

*Tons (US short) × 0.91 = metric ton

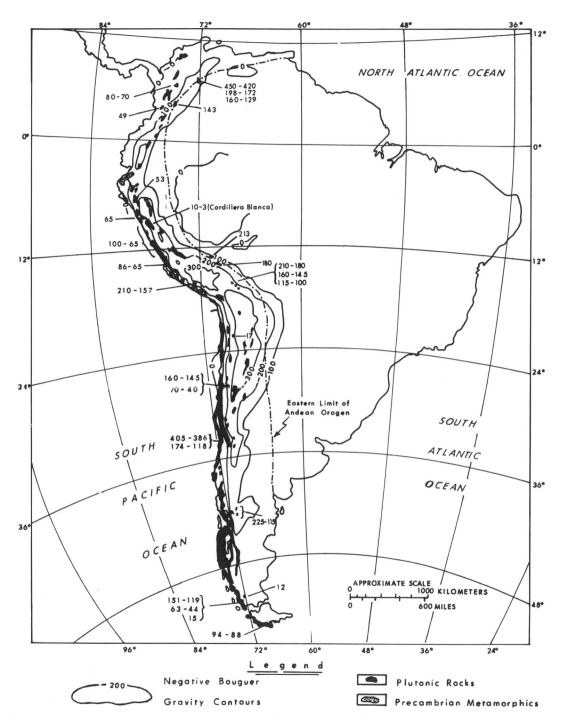

Fig. 2. Crustal elements of the Andean orogen. The Bouguer gravity anomaly map of the Andes is superimposed on the map, showing distribution of Precambrian metamorphics and the distribution of plutonic rocks. Radiometric ages are shown for the batholiths *(modified after Ericksen, 1975)*.

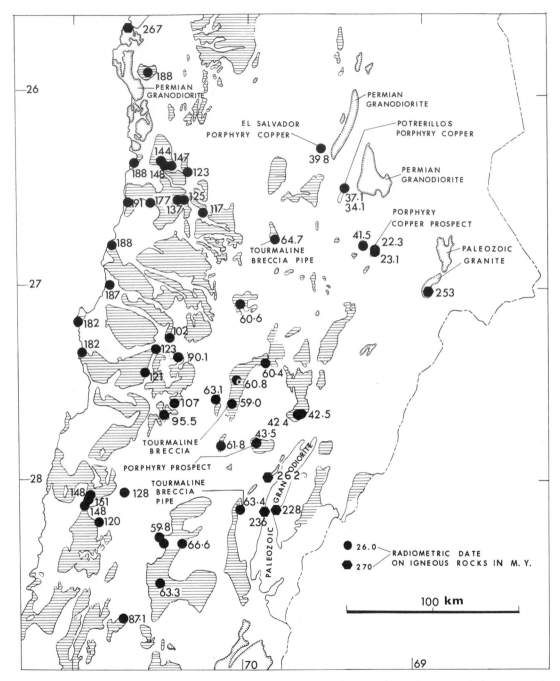

Fig. 3. Radiometric ages near El Salvador and Potrerillos, Chile. Radiometric ages of igneous rocks near El Salvador and Potrerillos deposits are typical of much of the porphyry copper province. Paleozoic or Precambrian igneous and metamorphic rocks underlie the deposits *(after Zentilli, 1974).*

Precambrian as well as Paleozoic crystalline rocks (Hollister and Bernstein, 1975). Here the Mesozoic sequence is less volcanic than in the El Salvador-Potrerillos area, although Mesozoic as well as Permian batholithic plutonism was widespread. Fig. 3 (after Zentilli, 1974) shows the known Paleozoic radiometric ages from plutons near the El Salvador and Potrerillos deposits. The general relationships depicted in Fig. 3 are typical of the porphyry province in much of the Andes; that is, an association of radiometrically dated Mesozoic arc-trench-batholithic complexes penetrating or overlying a pre-Mesozoic plutonic and metamorphic suite. Paleozoic batholiths, therefore, had in turn invaded an older sialic crust. Plutons with porphyry copper deposits penetrated older crust, including the Paleozoic batholiths, in many parts of the orogen.

It can be inferred by extrapolation of data through most of the Andean porphyry copper province that the mineralized plutons penetrated silicic crust that probably included Precambrian crystalline rocks and Paleozoic sediments, metasediments and plutons. Basic volcanic and ultrabasic rocks are rare in the pre-Mesozoic rocks. The general relationships of the older and younger rocks in the Arequipa area are shown in cross section in Fig. 4, modified from James (1971) and Ocola and Meyer (1973). This section, typical of most of the Andes where porphyry copper deposits are known, illustrates the older crust overlain by Mesozoic and Cenozoic rocks which may include arc-trench assemblages near porphyry deposits.

Where Precambrian rocks may be absent (Fig. 5), the continental crust intruded by the porphyry copper plutons may reasonably

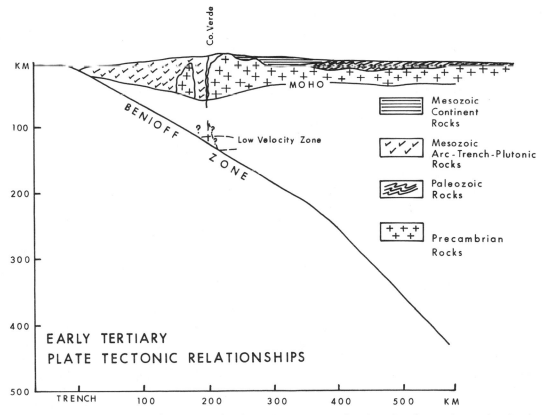

Fig. 4. Section through Arequipa, Peru. The Cerro Verde tourmaline breccia pipe and associated mineralization lie in the plane of the section through Arequipa, Peru. This section shows the hypothetical plate tectonic relations that existed in the Eocene at the time of mineralization (*modified after James, 1971*).

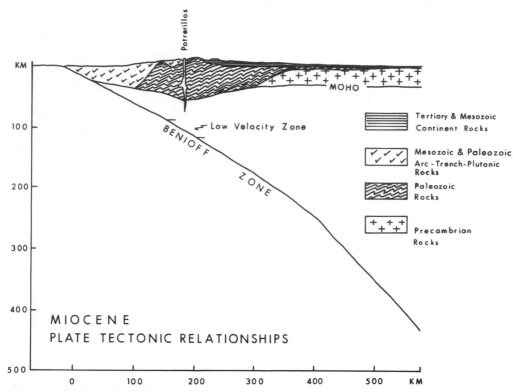

Fig. 5. Section through Potrerillos, Chile. Potrerillos is shown in this diagramatic cross section of the plate tectonic relationships as they existed in the Miocene at the time of mineralization.

be expected to include a thick section of metamorphosed Paleozoic rocks which, in effect, provide an analogous crustal environment to those areas where a Precambrian assemblage is present. In the porphyry copper province, the bulk of the crustal rocks consist of a pre-Mesozoic sialic assemblage that is believed to have an important influence on the nature and composition of the younger intrusions. The setting of the porphyries is therefore rather more cratonic than oceanic in character, if not strictly definable as cratonic.

Mesozoic and Tertiary: Mesozoic and Tertiary rocks are typically dominated by numerous, successive, arc-trench sequences near the coast and by continental volcanics and sediments in the interior. Both are intruded by plutonic rocks whose ages are penecontemporaneous with the arc-trench assemblages which Dickinson (1970), Hamilton (1969), and others have called sub-

duction-related. These are almost entirely quartz dioritic to granodioritic near the coast and in the areas where porphyry copper deposits are known. In places volcanic members of the arc-trench sequence are abnormally rich in copper and other metals (Goossens, 1973), and some volcanogenic massive sulfide deposits are known from the Mesozoic marine volcanic sequence of the Andes. This is in contrast to the pre-Mesozoic, where such deposits only rarely have been discovered.

Fig. 5 shows the general relationship of most of the Mesozoic arc-trench-batholithic rocks with porphyry prospects and deposits near El Salvador. The tourmaline breccia pipes and other important porphyry copper prospects have dates less than 70 m.y. (million years) in this area. The occurrence of the Permian and Mesozoic arc-trench-plutonic rocks to the west of the porphyry deposits is typical of other parts of the copper belt. Generally the pre-porphyry copper igneous assemblages consisting of arc-trench and associated plutonic rocks of Mesozoic age are close to and west of porphyry deposits in the Andes. Important exceptions to this simplification (Fig. 2) include deposits that lie entirely within the arc-trench rocks (as at Braden) or entirely west of and outside of the influence of the arc-trench-batholithic suite (as at Andacollo).

Although some members of the Mesozoic rock suites may contain an abnormal copper content, the setting of the deposits generally suggests that the copper content of the individual porphyry copper deposit is independent of the type of Mesozoic or pre-Mesozoic host rock. In any case, that part of the crust with the most anomalous copper content, the Mesozoic marine volcanic rocks, are only minimally involved in the Andean copper belt deposits.

The Tertiary section of both pre- and post-mineral age is largely andesitic, though Guest (1969) and others record local large volumes of dacite and rhyolite. Volcanic rocks coeval with mineralized plutons have been noted in some deposits (e.g., Pelambres [Sillitoe, 1973]), and in many of these examples, rhyolite and dacite are prominent. On the other hand, some mineralized plutons are not associated with observed coeval or comagmatic volcanic rocks (e.g., Copaquire [Hollister and Bernstein, 1975]).

In terms of genesis, the basic crustal setting for most Andean orogen porphyry copper deposits (Fig. 2) appears to have been similar to that of the Lowell and Guilbert (1970) model deposit of the southern Cordilleran orogen. A cratonic environment, which consists of a thick pre-Mesozoic crust (commonly including crystalline rocks of Proterozoic age), is associated with successive Mesozoic arc-trench-plutonic sequences that lie to the west of most of the deposits. No obvious source for the copper is visible or inferrable in the exposed or near surface portion of the crust.

Plate Tectonics: In this discussion, plates, transform, transverse, Benioff zone, and transcurrent faults as well as subduction zone are used in the sense of Herron's (1972) and Wollard's (1973) descriptive papers on the Pacific (including the Nazca) plate. The South American plate appears to be overriding the Nazca plate, which has, itself, spread from the East Pacific rise. Subduction begins with the trench and extends under the Andean orogen. Clague and Jarrard (1973) present evidence that the Pacific plate also had a generally northerly component of movement during the time when most Andean porphyry copper deposits formed. Therefore it would appear that both the suggested northerly trend of the Pacific plate and the easterly compressive force associated with subduction could have been operative during formation of most porphyry copper deposits.

For much of the porphyry copper province in the Andes, fold axes in Mesozoic and early Tertiary volcanic and sedimentary rocks parallel the coast and the present trench. It is assumed that paleo-trenches either coincided with or paralleled the present trench. A poorly defined belt of Mesozoic thrusting paralleling the trench has been found from west central Argentina to central Bolivia. Such structural features as folding and thrusting are omitted from Fig. 2 because they cannot be illustrated on a map with such a small scale.

The fold axes and thrust belt may have developed as a consequence of tectonic transport westward of the South American plate and from subduction at the Nazca-South American plate boundary. Most batholiths are also aligned parallel to the trench. If these quartz diorite-granodiorite batholiths are viewed as subduction-derived, then the Upper Paleozoic batholiths may imply that subduction preceded Mesozoic separation of South America from Africa, and that compression associated with subduction was initiated in the Upper Paleozoic. Should the dates on batholiths be correlated with rapid subduction, then indirectly radiometric dating of these rocks may be used to establish those periods when compression on an axis normal to the coast and trench were most actively present. In general, ages of batholiths tend to fall within periods of folding. Few porphyry copper deposits formed during the time of widespread batholith formation.

Some porphyry copper deposits are closely associated with right lateral strike-slip faults (e.g., Chaucha, Michiquillay, Copaquire, Chuquicamata). In most cases the faults parallel the coast and also therefore the fold structure, thrusts, and batholiths. They apparently reflect different tectonics from the subduction process that may have given rise to fold and thrust structures. Tertiary dates of porphyry copper intrusions that are closely associated with the dextral faults demonstrate that fault displacement took place during a tectonic regime separate from that which existed during the intrusion of batholithic masses. A north-trending component of movement on the Nazca plate may explain the movement on the dextral faults; these could be considered as manifestations of a megashear developed at and near plate boundaries as the Nazca plate moved north relative to the South American plate.

Evidence of dextral fault involvement in the development of these deposits is, however, rarely well documented. A speculation appears for the case of Michiquillay. The trend of the line of Tertiary stocks south of the Hualgayoc fault (Fig. 7) appears to have been the same as that for the axis of Tertiary intrusions north of the fault. If all the Tertiary stocks are the same age as Michiquillay (20.6 m.y.) and had originally occurred on one lineal trend, the offset on the fault since mineralization at Michiquillay has been 18 km.

Although some mineralization demonstrably took place on and near the strike-slip faults, it seems clear, from the evidence that led to the construction of Figs. 4 and 5, that a Benioff zone existed at the time of porphyry mineralization. The existence of a subduction zone may have facilitated the initiation and enriching in a zone of partial melting (a low velocity zone) of a porphyry copper system, but the strike-slip faults provided access and guidance into the upper crust for the porphyry in some specific deposits.

Hollister and Bernstein (1975) cite Copaquire, Chile, as an example of a porphyry copper deposit developed along a dextral fault, where mineralization and intrusion are associated in space and time with a right-lateral strike-slip fault that had pre-, intra-, and post-mineral movements. Evidence cited by James (1971) suggests that a trench and subduction zone existed at the time of mineralization (22 m.y.). The dextral fault evidently was active during the time when the subduction zone existed, and both phenomena may have been important to the development of the deposit. A similar argument may be established for Chuquicamata (south of Copaquire), because the structural setting of the two deposits is similar.

Further speculations have been made by Sillitoe (1972) concerning the derivation of the porphyry magma and metal from a subduction zone. Figs. 4 and 5 show the subduction zone as it may have existed in several parts of the porphyry copper province at the time of metallization. The parallelism between the line of higher grade deposits and a probable trench, plus the arguments for a subduction zone associated with the trench (James, 1971), make the Sillitoe (1972) hypothesis a plausible speculation.

Fig. 4 gives the plate tectonic relationships that may have existed in the earliest Tertiary when the Cerro Verde-Santa Rosa deposit formed. If the section is correct, then the deposit formed about 150 km vertically over the Benioff zone. Similar depth estimates have been made for Chaucha (Goossens and Hollister, 1973) and Michiquillay (Hollister and Sirvas, 1974).

The Andean copper belt lies over the projected downdip position of a paleo-subduction zone where it would have existed at depths of 100 km to 150 km below the surface. A zone of partial melting is most commonly proposed within these limits. Porphyry copper deposits east and west of the Andean copper belt exist over the paleo-subduction zone where this zone would project to either shallower or deeper penetrations of the mantle, outside of the most commonly proposed region of partial melting. Coincidentally, porphyry copper deposits outside of the Andean copper belt are smaller and contain a lower copper grade than are found for deposits within the belt.

Geology of the Deposits

On the basis of their internal structure, South American porphyry copper deposits can be separated (after Ruiz, 1965, and Hollister, 1974) into either a tourmaline breccia type or a stockwork type. Tourmaline breccia type deposits are summarized in Table 2. They may have no obvious association with regional structure, or with fault or fold structures related to failure in a moving plate, but potassium-argon dating does support the contention that these deposits formed simultaneously with movement of an oceanic plate down a Benioff zone.

The stockwork-type deposits summarized in Table 1 contain intersecting veins and veinlets and may more frequently be associated with major regional faults. The stockworks may be genetically related to displacement on major strike-slip faults or some other structural elements evolved in the margin of a moving plate.

The following sections present some characteristics and modes of occurrence of the two structural types of deposit.

Petrography: The petrography of intrusions closely related to ore in time and space, in either structural type of deposit and regardless of age or location, seems remarkably constant. Similar types of alteration are ubiquitous in the mineralized intrusives, and these tend to mask any primary petrographic differences. Despite this, megascopic examination provides little reason to suspect that substantial compositional differences exist between dacite at Toquepala, quartz monzonite porphyry at Michiquillay, quartz monzonite porphyry at Campana Mahuida, or dacite at El Teniente. These represent a considerable range of ages (from 4.3 to 58.7 m.y.), as well as variable distances from the leading edge of the continental plate. Minor compositional variations from the dominant quartz monzonite porphyry do exist; for example, the granodiorite at El Salvador and the quartz diorite at Mocha. Variants display a random distribution within the plate; that is, they occur at erratic distances from the trench.

Limited data on the petrography of the intrusions listed in Tables 1 and 2 do not permit construction of plots of various parameters on individual deposits, as in Figs. 36a, b and 42 (pp. 103, 104, 130) for the Cordilleran orogen. However, the petrography of Andean porphyry copper deposits may be generally described in terms of relevance to their genesis. All intrusions associated with the major porphyry copper deposits appear to be consistent with the model developed by Lowell and Guilbert (1970).

Interpretation of chemical data from Haynes (1972) and Palacio and Oyarzun (1975) tends to support the analysis based on reconnaissance petrographic identifications, at least in the area from El Salvador to the east. Fig. 6, modified from Haynes (1972), appears to show a nearly constant potash index for sulfide-associated, intermediate, porphyritic intrusions irrespective of distance inland from the trench for the area studied.

Lacy (1957, p. 3) gives one explanation for the lack of strong variance, suggesting that the rocks are a common end product of differentiation. The greatest difficulty with this concept, however, is the absence, in some deposits, of the products of differentiation other than the mineralized intrusion. In the extreme case, e.g., Michiquillay (Hollister and Sirvas, 1974), the only igneous rock exposed is the intrusion that hosts mineralization. At the surface near Michiquillay (Fig. 9), only the mineralized intrusion or other small intrusions of nearly the same composition are found near the ore deposit. Volcanic rocks at a distance from Michiquillay appear to have their own source vents.

If similarity of the crustal section cut by these various Andean mineralized plutons had influenced their composition, then the small range in chemical composition apparently present for the intramineral intrusion may reflect the similarity of the crustal setting that hosts the deposit. This hypothesis is a speculation not amenable to proof.

Dickinson (1970) presents evidence that most melts in continental-type orogen are derived from the upper side of the Benioff zone. Gilluly (1971) gives an even more convincing theoretical explanation for ande-

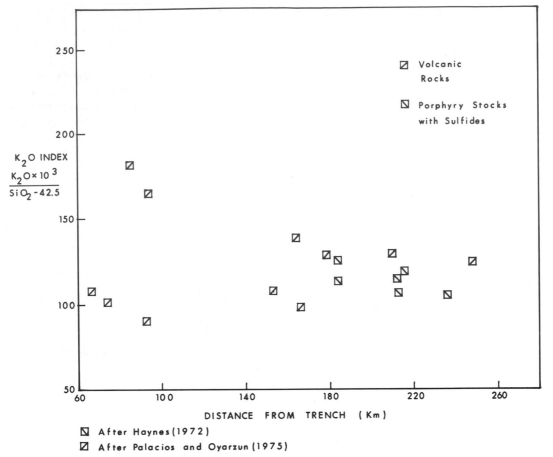

Fig. 6. Potash index of igneous rocks, Atacama Region, Chile. Mineralized plutonic rocks appear to have potash index trends similar to those of extrusives for that segment of the orogen where porphyry copper deposits occur *(modified after Haynes, 1972).*

site and granodiorite formation from the sub-duction zone. Fig. 6 illustrates data compatible with evidence used by each for calc-alkalic magmas. Generally, as distance is gained from the trench, the potash content of those igneous rocks derived from the subduction zone tends to increase. Apparently intrusions close to ore in age and space in Andean porphyry copper deposits may make an important exception, because the potash content of these specific intrusions has not been shown to have any systematic variation away from the plate's leading edge. A comparable analysis could be made in the Cordilleran orogen for intrusions close to ore in time and space from Yerington generally east through Battle Mountain, the Robinson dis-

trict, and Bingham. Here again, these mineralized intrusions do not appear to heed the rule that subduction-derived magmas generally have a significant increase in potassium and alkalies eastward from the plate edge.

James (1971), Clark (1970), Haynes (1972), and Farrar (1970) each make a case for a general eastward migration of igneous activity with time in the Andean orogen. Age dating on porphyry copper deposits alone, to the contrary, shows an almost complete time independence in their occurrence geographically within the plate. Paramillos Sur, on geologic bias, has been listed as the oldest of any deposits shown on Tables 1 and 2 (Ljungren, 1970) and it lies in the eastern foothills. El Teniente (Clark, 1972)

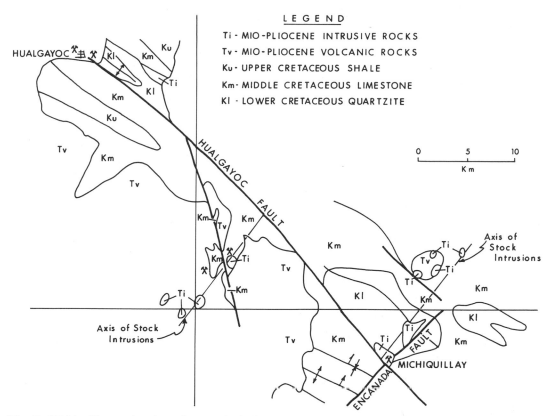

Fig. 7. Michiquillay regional geology. Michiquillay occurs on the intersection of the Encañada and Hualgayoc faults. It is conceivable that the northeast-trending line of stocks may have formed at the approximate time of development of Michiquillay, since the axes of the two lines of stocks are subparallel and stocks on either side of the Hualgayoc fault are petrographically similar. If this is the case, right lateral movement may have occurred on the Hualgayoc fault after the stocks were intruded. Recent movement has also occurred on Encañada fault, however, since it is visible on both sides of the Hualgayoc fault (*modified after Benavides, 1956*).

appears to be the youngest, and it lies west of the divide. The pre-Karoo Haib porphyry copper deposit in South-West Africa probably formed prior to the separation of South America from Africa, and it lies even further east than Paramillos Sur. If potassium-argon (K-Ar) dating is to be believed, the erratic distribution of porphyry copper deposits in time and space in the Andean orogen needs some fuller explanation than is now offered by Sillitoe's (1972) plate tectonic model.

Herron (1972), Le Pichon and Hayes (1971), and others have shown that both paleomagnetic evidence and an assumed 2.0 cm/yr spreading rate calculated for the Atlantic coincide to date the initial separation

of South America from Africa at about 140 m.y. Separation appears to have been completed over the next 35 m.y. Some magmatic activity and metallization within and near the copper belt preceded this event, because 180 to 199 m.y. intrusions are found in Bolivia (Lohmann, 1970, and Clark and Farrar, 1973). Tin mineralization has been associated with a few of these magmas. Clark (1972) has demonstrated the existence in Chile of Permian (228 to 261 m.y.) and lower Mesozoic (177 to 191 m.y.) granodioritic batholithic intrusions, and other data (Fig. 3, modified from Zentilli, 1974) corroborates this conclusion. Triassic continental volcanics are known within the Andean orogen. Epizonal magmatism (and also porphyry copper-type deposits) therefore may logically date from the Permian if porphyry copper formation may be related in any way with magmatic events of the type dated. Comments by Sillitoe (1972) and others on the copper content and depth of erosion, on the other hand, suggest that the younger deposits (Tertiary) are more likely to be economically significant. It is clear that the separation of South America from Africa is inconsequential in the history of porphyry copper evolution in the Andean orogen, although the most intensely metallized known deposits date from well after that event.

Alteration: Alteration phenomena in each Andean porphyry copper center appear to fall within the zones compiled by Lowell and Guilbert (1970) and others for alteration of quartz monzonitic to quartz dioritic composition in intrusion. All major porphyries listed in Tables 1 and 2 have a phyllic zone, surrounded by an argillic zone, and an outer propylitic zone. The younger, less intensely eroded deposits may not, however, have an exposed central potassic core zone found in some of the older deposits. In some younger deposits extensive biotization (as in Braden) may represent potassic alteration, but orthoclase may not be well developed in these areas. Each of the deposits contains the typical pyrite halo, as described by Rose (1970), around the copper-bearing center. This is common to all deposits, regardless of the presence or absence of other alteration mineral assemblages.

The phyllic zone generally appears to be coincident with the areas of greatest fracturing, and along with the potassic zone is usually the locus of copper mineralization where metal zoning exists. In some deposits (e.g., Chuquicamata and Michiquillay) copper occurs most prominently in the phyllic zone, while in others (e.g., San Salvador, Gustafson and Hunt, 1975) copper is well developed in a potassic zone. The alteration zones are much the same as at Butte (Sales and Meyer, 1948).

In the phyllic zone, the quartz, sericite, pyrophyllite (if present), other silicates, and pyrite are systematically or zonally developed from fracture surfaces into the host rock. The suites of alteration minerals change outwardly, on a large scale, from the potassic core through the phyllic, argillic, to the propylitic zones; but mineral zoning away from fractures within any one zone may display the entire range of alteration minerals. Where the host rock of the deposit is cut by many closely spaced intersecting veins and veinlets, individually developed, zonally arranged mineral assemblages overlap; and the cumulative or aggregate effect provides the mineralogy found in one or the other of the respective alteration zones. Therefore, quartz, sericite, and the clay minerals are not distributed in a random pattern in the phyllic zone. Rather they are zonally developed about individual fractures and the alteration designation is determined by the pervasive dominance of key minerals.

In the breccia pipe model, repeated shattering and rotation of fragments in the conduit during the passage of hydrothermal fluids may permit uniform alteration of the entire pipe, which is enveloped by halos of lower grade alteration.

In most Andean deposits, regardless of structural type, alteration mineralogy has been inadequately described in the literature. Where tourmaline occurs as an alteration product, it may appear as an added silicate mineral in any of the zones but most commonly is found as a replacement for original ferromagnesian silicates. Pyrophyllite and other minerals megascopically indistinguishable from sericite may occur in the phyllic zone (as at El Salvador and Chuquicamata), but these are easily confused with sericite and are only rarely reported.

Structure: The localization of South American porphyry deposits appears to be a function in some manner of internal failure within the continental plate. Wollard (1973) and others have inferred extensional conditions above the subduction zone at the margin of the continental plate. All porphyry copper deposits apparently formed in an environment of tensional stress, as most copper and molybdenum sulfides occur as fracture filling rather than wall rock replacement or dissemination.

In breccia pipes, the rotated fragments are cemented with gangue and metallic sulfide minerals. Stockworks are composed of many, repeatedly opened veins and veinlets filled with virtually the same sulfide mineral suite and in much the same paragenetic sequence as found in the breccia. In the breccia pipes the release of pressure that led to the original piercement helped generate the extensional conditions that produced the internal structures of the pipe. For the stockwork model, some peculiarity of the tectonic environment led to local extensional conditions. The causative factors may be unique for each stockwork deposit. Fracture patterns of individual deposits may betray the genesis of the stress-strain relationships that led to fracture development.

Hollister (1974) cites evidence that many stockwork deposits are associated with strike-slip faults. Specific stockwork porphyry copper deposits can be related genetically to regional structural features. Incomplete coverage by large-scale published geologic maps prohibits assessment of tectonic relationships for all stockwork deposits; however, where regional structures are known, a genetic relationship between major faults and stockwork deposits is frequently demonstrable. Because few tourmaline breccia deposits are close to or appear to be directly influenced by major faults, no such relationship may be claimed for them.

To avoid confusion in interpretation of the two structural models, it should be noted that weak tourmaline alteration and small breccia pipes may occur erratically in the stockwork deposits. However, the stockwork deposit rarely has dominant circular fracture or veinlet patterns, although it may have an identifiable weakly developed circular pattern about a specific stock. On the other hand, breccia pipes tend to have strong circular and radiating fracture (veinlet) patterns around a central pipe. Either type of fracture pattern may form a well developed stockwork of veinlets. The diagnostic fracture pattern of breccia pipes is circular, whereas the characteristic pattern of stockworks is formed by sets of conjugate veinlets, or by intersecting but offsetting sets of parallel veinlets.

Plates 2 and 3 (pp. 22A, 22B) are photographs of stockworks.

Examples of Stockwork Types: Where plate failure appears to be a major transverse or transcurrent fault or other strike-slip structural element developed in the plate, stockworks may have formed directly in subsidiary fractures conjugate to regional faults. Because transcurrent and transverse faults are the most important and persistent ruptures developed as a consequence of plate movement, association of some porphyry copper deposits with those faults might be expected. Fractures conjugate to major faults may have provided openings for hydrothermal fluids, but the major regional faults remain barren.

Chuquicamata is perhaps the best example of this type. The West Fissure, which strikes N10E, appears to have acted within the parameters Herron (1972) used for regional transcurrent faults. Perry (1952) states that "the West Fissure was an early, pre-porphyry structure along which profound, primary crustal adjustment occurred. The evidence is strong for believing that it controlled emplacement of the porphyry, quartz and sulfides, ending its activity in final relief of stresses after the close of the mineralizing epoch." Accompanying this fault and appearing on its projection for 150 km, Ridge (1972) finds "a long narrow belt of stocks . . . that extends from Caracoles to El Abra in the north." At the time of porphyry mineralization (29 m.y.) Clague and Jarrard (1973) imply a northerly component of movement of the Pacific plate (interval 20-42 m.y.). Their hypothesis suggests that the Nazca plate moved north relative to South America; and the movement on the West Fissure, which parallels the plate boundary, is on the west side of north (Fig. 8).

Perry (1952) further states, "The intensely mineralized belt constituting the ore body contains veins and innumerable criss-cross vein structures with intense mineralization and alteration of intervening porphyry." Perry (1973) points out that the numerous large veins that accompany the veinlets within the ore deposit are incompatible with a strict definition of stockwork. The term stock-

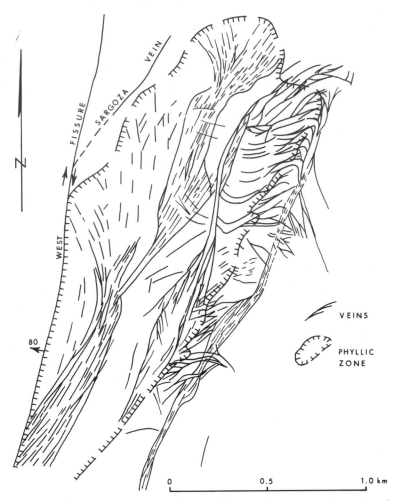

Fig. 8. Structure at Chuquicamata, Chile. The Chuquicamata deposit is a vein and stockwork filling of ore and gangue minerals on the east side of a regional fault. The fractures are those that should be expected as right lateral displacement occurred along the West Fissure. The conjugate set developed are typical of this type of stockwork deposit (*modified after Lopez, 1942*).

work used to describe Chuquicamata mineralization is, however, broadly accepted.

Lopez (1942) indicated that mineralized fractures are a conjugate set genetically related to the West Fissure, and that they formed when the west block of that fault moved north relative to east. Fig. 8 modified from Lopez (1942) shows the relationship of phyllic alteration zone to mineralized fractures. Repeated opening of stockwork fractures controlled hydrothermal activity, including alteration and sulfide deposition.

Where transcurrent faults intersect major strike-slip faults, simultaneous movement on each fault may permit a stockwork to develop at the intersection. In such stockwork deposits the fracture systems commonly appear in and near intrusions emplaced and controlled by intersections. Michiquillay lies at the intersection of the N45W striking Hualgayoc fault, and a N50E strike-slip crossing structure (Fig. 3). The Hualgayoc fault has had largely horizontal movement, but doming related to intrusion gives the impression of local vertical movement. Both strike-slip trends as well as conjugate frac-

Plate 1. Tourmaline breccia at Toquepala. Irregular breccia fragments are cemented with a black tourmaline-rich matrix. Quartz and sulfides occur with the tourmaline.

Plate 2. Stockwork outcrop. Veins and veinlets of quartz-limonite occur which cut sericitized andesitic volcanics in this leached capping over ore. At depth the vein structures contain hypogene chalcopyrite and pyrite with quartz.

Plate 3. Stockwork outcrop. Near surface mine bench at Esperanza mine showing vein and veinlet structures emphasized by iron oxide content. This leached capping is over supergene and hypogene ore.

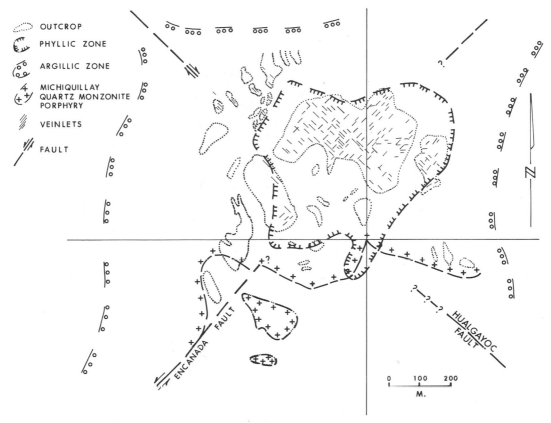

Fig. 9. Surface geologic map, Michiquillay. Intersection of the Hualgayoc and Encañada faults is the center of alteration and mineralization for the Michiquillay deposit. The stockwork developed is largely derived by repeated simultaneous displacements along both faults.

tures are developed in the stockwork. As can be seen in Fig. 9, the resulting network of veins and veinlets is fully as striking a feature as the Chuquicamata stockwork, but the differences in orientation of the veins and veinlets is clear. Fig. 9 shows the relationship between stockwork and hypogene alteration zones at Michiquillay. The stockwork was produced by simultaneous movement of both faults, and the area of intersection is clearly the center of hydrothermal activity. In this instance, one component of the stockwork is itself part of the transcurrent fault.

Where major transverse faults cut the leading edge of the continental plate, porphyry copper deposits may develop in situations somewhat analogous to those examples cited previously. The Chaucha deposit appears related to such a major structure identified with plate motions. Goossens (1972) points out that the deposit is on the Chaucha fault, a strike-slip structure that Goossens and Hollister (1973) define as a transverse fault having more than 8 km of post-Cretaceous horizontal displacement. The Chaucha porphyry copper occurs at the intersection of the Chaucha fault and the northerly trending Cordillera fault. Stockworks in this deposit have dominant trends N10E and E-W (Fig. 10), reflecting regional faults; and the center of intrusive activity, hydrothermal alteration, and mineralization is the intersection of the two faults. In this deposit, faulting, intrusion, and mineralization are demonstrated to be a continuous sequence of events.

Where veinlets and veins form through stress release in response to intrusion of a circular stock, the veinlets may describe a

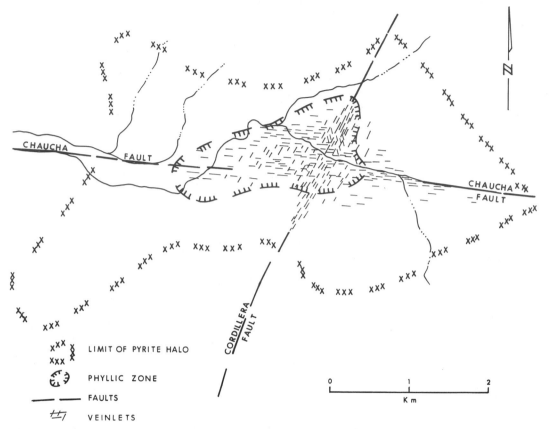

Fig. 10. Surface geologic map, Chaucha. Intersection of the Chaucha and Cordillera faults is the center of alteration and mineralization for Chaucha. The stockwork is largely composed of closely spaced mineralized segments of each fault that have developed during simultaneous displacements along both *(modified from Ljungren, 1970)*.

circular pattern. The circular pattern may be superimposed on, or be visible through, the normal stockwork trend developed during formation of the deposit. Campana Mahuida, Argentina (Fig. 11), exemplifies circular fractures occupied by veinlets of quartz and sulfide that cut Cretaceous sandstone. They are related to a stock that does not outcrop. The first hole drilled in the United Nations program at Campana Mahuida penetrated the center of the circular pattern and intersected an altered and mineralized intrusion. Veinlets trending N35E are superimposed on the circular set, and some dikes of altered igneous rock follow this alignment. Intrusion of the stock, formation of the circular and northeast sets, and alteration and mineraliza-

tion have been demonstrated to be part of one continuous process (Ljungren, 1970).

Although strike-slip movement seems most common on large regional faults with a demonstrable genetic tie to stockwork porphyries, high-angle deep-penetrating dip-slip normal and reverse faults also occur in the continental plate and may also be significant in porphyry copper development. Morococha (the Toro Mocho porphyry copper) may be an example of a deposit that developed in conjunction with major faults having prominent dip-slip movement. Terrones (1958) infers that the N25W-trending regional faults present there (displaying dip-slip movement) appear to have influenced early, premineral, epizonal magmatism. Eyzaguirre, et al.

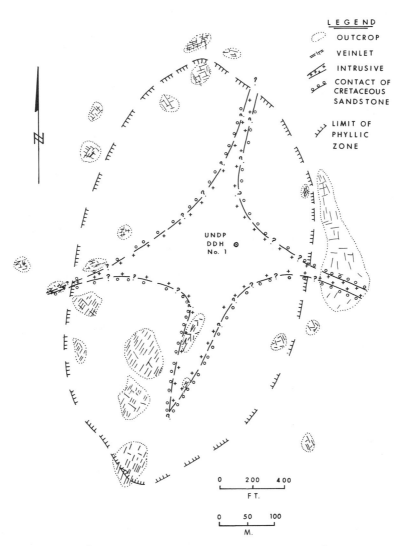

LEGEND

OUTCROP

VEINLET

INTRUSIVE

CONTACT OF
CRETACEOUS
SANDSTONE

LIMIT OF
PHYLLIC
ZONE

UNDP
DDH
No. 1

```
0      200    400
        F T.

0      50    100
         M.
```

Fig. 11. Structure in outcrop at Campana Mahuida, Argentina. Interpretation of structure and alteration are shown for Campana Mahuida deposit, based on outcrop studies. A circular set of fractures appears to be annular to a nonoutcropping plug. Dikes radiating from the quartz monzonite plug may be seen in outcrop. The plug is the central heat source for mineralization in the district (*modified from Ljungren, 1970*).

(1975), show that the sequence of plutonic and hydrothermal activity may have spanned only 1.1 m.y. However, the porphyry copper deposit in Morococha was explored after comprehensive descriptions of the district were published, and a link between mineralization and the faults has not been demonstrated. Dominantly east-west fracturing is present in what is now known to be the porphyry copper center (Haapala, 1949),

although these faults could be subsidiary fractures. The Haapala (1949) discussion of breccia at Morococha illustrates the occurrence of breccia pipes in a stockwork deposit. Cerro de Pasco Corp. maps (Fig. 12) do not clearly show the N25W structures referred to by Terrones (1958).

Examples of Breccia Pipe Type: In contrast to the stockwork deposits, Ruiz (1965), Sillitoe and Sawkins (1971), and Hollister

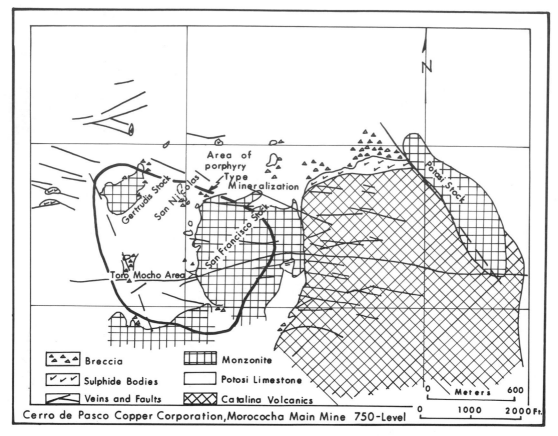

Fig. 12. Morococha, Peru, including Toro Mocho area. An outline geologic map of Morococha shows that the area which may include porphyry-type mineralization includes both quartz monzonite and intruded limestone. Skarn development is widespread in limestone near igneous contacts (*modified after Haapala, 1949*).

(1974) imply a separate model for tourmaline breccia pipes. This discussion generally follows the Sillitoe and Sawkins (1971) classification without change, but expands it to include porphyry copper deposits in the Andean orogen they did not mention. On a regional basis, the outstanding feature of tourmaline breccias is an apparent lack of dependency on or involvement with a regional fault. This applies to Toquepala (Richard and Courtright, 1958) and to El Teniente (Howell and Molloy, 1960). Similarly, no genetic dependence on regional structure can be discerned from published detailed maps of Disputada, Los Pelambres, or Cerro Verde.

Plate 1 (p. 22A) is a photograph of tourmaline breccia from Toquepala.

The controlling structure for nearly all subsidiary features of the breccia pipe type is the pipe itself or the intrusion that may have preceded it. In South America the common association of tourmaline with breccia pipes and the tendency for pipes to occur near the leading edge of the moving continental plate are characteristic. Tremendous volumes of boron needed to fill the pipes have no easily discernable source in the subduction zone. The generally circular or elliptical form of the pipes, their circular and radiating fracture patterns, their steep or vertical tubular shape, and their random distribution may best be explained by strong vertical movement of material that punctured the crust at will.

Diagnostic structural features of the large breccia pipe model usually include the following:

1) A pipe composed of angular fragments cemented with tourmaline and other minerals.

2) Common occurrence of circular and radial mineralized fractures around the tourmaline breccia pipe.

3) A pebble breccia (e.g., Braden pipe) cutting both 1 and 2.

Apparently the fracturing was initiated by an intrusion, or by boron-rich hydrothermal fluids, or both, but evolution of structures now visible in and around some pipes required repeated movements within the conduit. These movements may have been of both collapse and injection type, and they led to fracturing of the walls, as well as to repeated brecciation in the pipe. Involvement of magma with the earliest fractures can be demonstrated in a number of deposits. Dacite at Toquepala has a close time-space relationship to mineralization and occupies an annular ring. Howell and Molloy (1960) describe a similar dacite cone sheet at El Teniente, which they state is "closely associated in age with ore." Paragenetically, copper is penecontemporaneous with tourmaline in the majority of these deposits, although metals other than copper may accompany the tourmaline. Additionally, some pipes are barren of sulfides.

The general setting of Toquepala (Fig. 13) and detail of a portion of the pipe (Fig. 14) illustrate a typical breccia pipe deposit. Veinlets encircle the tourmaline breccia or, more rarely, are radial to it. In detail the breccia irregularly cuts the peripheral stockwork of veinlets. Although some veinlets cut the breccia pipe, the repeated veining suggests a more complicated relationship between the two. Ore grade mineralization lies in both the breccia and the circumferential stockwork of veinlets in the adjacent wall rock. In the Toquepala ore body, the stockwork of veinlets around the breccia is not as dense as that at Chuquicamata, although some ore at both Cerro Verde and El Teniente extends into the stockwork zone. In Toquepala, however, as with most breccia pipe types, the best grade occurs in the breccia, with the hypo-

gene grade weakening away from it. Intensity of fracturing and quantities of hypogene copper and molybdenum sulfide mineralization fade as distance is gained from the pipe. Dacite fills a few of these circular peripheral fractures, but it in turn is the host for ore minerals occurring in circular veinlets. Minor amounts of ore minerals may be found disseminated in the dacite as well. The phyllic alteration zone occupies the pipe and the adjacent walls where fracture density is greatest.

Fig. 14 demonstrates a feature found locally in some large pipes. In some large

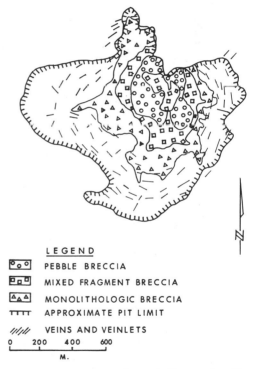

LEGEND

⬛ PEBBLE BRECCIA

⬛ MIXED FRAGMENT BRECCIA

⬛ MONOLITHOLOGIC BRECCIA

TTTT APPROXIMATE PIT LIMIT

///// VEINS AND VEINLETS

0   200   400   600
M.

Fig. 13. Structural outline of Toquepala, Peru. Toquepala tourmaline breccia (ore breccia) is ringed by an annular mineralized fracture pattern. Annular fractures may have formed partly in response to intrusion of a pre- or early mineral dacite plug, but strong mineralization within these fractures indicates some formed in response to development of ore breccia. Pebble breccia is post-sulfide but has appreciable sulfide content since numerous rounded fragments of tourmaline breccia are incorporated in it. It represents the latest hydrothermal phase.

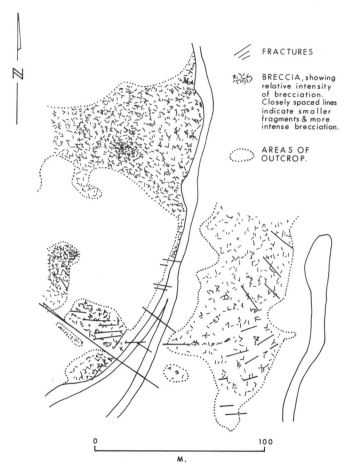

⟋⟋ FRACTURES

✴✴✴ BRECCIA, showing
relative intensity
of brecciation.
Closely spaced lines
indicate smaller
fragments & more
intense brecciation.

⋯⋯ AREAS OF
OUTCROP.

0   100
                M.

Fig. 14. Detailed surface geologic map of Toquepala. Details of breccia development in outcrop are diagramatically portrayed by intensity of "chicken" scratches. Smaller breccia fragments more closely packed together are shown by shorter, more clustered scratches. Fractures cutting breccia are intra-mineral in age and are part of the circular fracture pattern.

pipes within any small area, individual breccia fragments may tend to have about the same mass, and fines are missing. Where mixing is thorough, and the breccia consists of numerous fragments of rock (including both dacite and tourmaline breccia itself brecciated), size classification is best developed. Change in the size of fragments from one part of a mixed fragment breccia to another may be gradual, and the change from a mixed fragment breccia to wall rock nearly always occurs in a breccia zone containing fragments of wall rock only. One possible explanation of the size classification phenomena is that fines were carried upward and away by rising fluids, while coarser fragments were sorted according to mass. Fines may also have reacted with fluids to form tourmaline and other gangue silicate minerals. Variation in fragment size from place to place may be seen in Fig. 14 as nonsystematic but gradual. The change from a mixed fragment breccia to unshattered wall always involves passing through a zone of breccia containing only fragments of the rock found in the wall.

Where a number of small diameter pipes have penetrated the surface (e.g., Cabeza de Vaca, Chile), fracture patterns may be complex due to overlapping. Segerstrom

(1967) notes "90 breccia bodies which are either circular or lenticular in plan in an area 1300 by 2000 meters" in this area. Thus the example presented for Toquepala may be the simplest case for a single large pipe, with the possibility remaining that numerous closely spaced pipes may have interlocking boundary fracture zones.

Post-tourmaline breccia (e.g., pebble breccias) and igneous rock also exist in this structural model, indicating continued igneous and hydrothermal activity after the main episode of tourmalinization. Later breccia also may be devoid of any clear connection with regional structures at the surface and may be mineralized with sulfides but not with tourmaline in typical stockwork-type deposits (as at the Chuntacala ore body at Cuajone).

DISCUSSION

Unifying features of porphyry copper deposits in the Andean orogen are similar petrography, similar alteration zoning in and about ore, and nearly identical paragenesis of sulfide minerals regardless of structural type. Each structural type (stockwork or breccia pipe), however, displays individual details of ore emplacement, and the stockwork model tends to be more intimately involved with regional faults whose movements are compatible with plate tectonics. Deep regional faults may have intersected the source of magmatic-hydrothermal fluids and given rise to stockwork-type deposits during repeated movement that caused tensional openings in intersecting or conjugate sets. Lacking this escape, hydrostatic pressure in the fluids could build up to the point where a piercement structure and breccia pipe formed. Thus a common source but different mechanism of emplacement may explain many similarities of the two deposit types. Boron in breccia pipes may signify intense pneumatolitic activity in breccia pipe deposits, in contrast to stockworks in which tourmaline is rare.

Most known K-Ar dates for larger deposits in the Andean orogen are listed in Tables 1 and 2. Not shown are data for smaller porphyry deposits such as that at Cerro Rico, Argentina, with a date of 5.9 m.y. Additional dates are known for incompletely described porphyry deposits, but these tend to fall within the range of those listed and do not invalidate generalizations derived from the data given.

Misleading results from isotopic dating may have a variety of causes. It is assumed that those using the data are aware of such limitations, and qualifications applied to ages derived from the Andean orogen will not be examined.

Clark (1972) provides a 4.32 m.y. age for El Teniente, and for this deposit Ridge (1972, p. 593) speculates, "Tennantite mineralization took place in the Pliocene." The geologic evidence supports the K-Ar age. Lowell (1974) provides a date of 4.6(?) m.y. for Rio Blanco and of 10 m.y. for Los Pelambres. Goossens (1972) mentions two young-age determinations for Chaucha. The 9.9 m.y. age is preferred as the more plausible geologically. Michiquillay also has two ages recorded, but Damon's (1973) explanation for the 20.6 m.y. date is sufficient to justify its use. A. H. Clark (1976) advises that the 22 m. y. date noted in Table 1 for Copaquire disagrees with a date determined at Queens University. He has obtained a 30 m.y. date for this deposit. Dates presented in the tables generally have some geologic support and therefore are believed to be valid.

The narrow range in time for K-Ar dates on Andean porphyry copper deposits contrasts to the broader spacing of dates reported for similar deposits of the Cordilleran orogen.

A wider age spread for Andes deposits seems possible when more of them are investigated and brought into production. With present information, however, youth is apparently a general characteristic of deposits throughout the length of the Andean orogen, from Pantanos south to Campana Mahuida.

Published information on ore mineralogy exists for some Andean deposits. Copper sulfosalts (generally enargite and tennantite) are more common in Andean, as opposed to Cordilleran orogen deposits. Enargite appears to be an important constituent of some ores at Morococha and Chuquicamata and the dominant copper mineral in Cerro de Pasco and some other Andean deposits. Most published mineralogic descriptions are too incomplete, however, to permit broad generalizations for the entire Andean porphyry copper province. Chalcopyrite remains the most common hypogene copper ore mineral, and most deposits contain molybdenite and minerals of zinc, lead, and silver in some type of zonal arrangement with copper. Metallic sulfides occur preferentially as fracture fillings of tectonically formed openings.

Ore reserves data in the two tables derive from the literature, with sources listed in the bibliography. Some reserves seem substantially higher in grade than are reported for similar deposits exploited in the Cordilleran orogen. High grade ore is usually enveloped by halos of progressively lower grade material; the copper content fades as distance from the mineralizing center increases. For this reason quoted reserves do not reflect ultimate tonnage or grade potential, and mining may involve a much larger tonnage of lower grade material. Lowell (1974) cites 1400 million mt reserve at Chuquicamata with a grade of 1.2% Cu, but an additional 1300 million mt of 0.3% Cu also exist. He cites an additional 4300 million mt of 0.3% Cu around the El Teniente (Barden) reserves, which are given in Table 2 as 4000 million mt of 1.05% Cu.

Data on past mining may be similarly misleading. V. D. Perry (1973) stated that past tonnage and grade of ore treated at Chuquicamata "to 1952 as 363 million tons with a recovered grade of 1.4%." Waste moved in this operation averaged 0.4% Cu and would have been commercial for mines in much of the Cordilleran orogen; it should be included with produced ore for comparison with other mines. Unfortunately such data are missing for most Andean mines.

CONCLUSION

The nature of porphyry copper deposits of the Andean orogen permit the structural geologist to establish distinct stockwork and breccia pipe models. Presence of tourmaline preferentially in breccia pipes lends support to the distinction. In other porphyry copper provinces where tourmaline may appear more rarely, establishment of the two models is less easily recognized and less universally accepted, but may be real, nevertheless.

The erratic geographic distribution of Andean mineralized calc-alkalic intrusions through time finds no easy explanation in the plate tectonic evolution of South America. Granitic intrusions generally are oldest closest to the trench and younger to the east. Porphyry copper magmas are haphazard in time and spatial distribution, however. Variations in composition from quartz monzonite may occur in mineralized intrusions, but these display no systematic pattern as a function of distance from the trench. In particular they do not become more potassic toward the east.

Thus time, space, and compositional factors of porphyry intrusion do not conform to the general rule of being younger to the east for subduction-derived magmas. Granitic intrusions do have this tendency.

Independence from both age and compositional trends of subduction-generated magmas shown by mineralized intrusions suggests an origin distinct from the typical island arc-batholith-associated magmas.

On the other hand, most larger porphyry copper deposits lie within the Andean copper belt, which in turn coincides with the projected surface trace of 100 to 150 km depth interval cut by paleo-Benioff zones existing at the time of mineralization. The zone of partial melting (low velocity zone) under modern conditions most frequently is suggested to be somewhere within this same depth interval.

Existence of both stockwork and breccia pipe models implies that a porphyry copper system generated at depth may be guided to the surface by large regional structures. In their absence, the hydrostatic and lithostatic pressure within the system is adequate to bring it into the upper crust as a piercement, which may then develop as a breccia pipe.

REFERENCES AND BIBLIOGRAPHY

Anon., 1970, "Peru's Expanding Copper Role," *Mining Journal*, Jan. 9, p. 25.

Anon., 1971, "Chile," *World Mining*, Aug., p. 80.

Bellido, E., and Simons, F. S., 1957, "Memoria Explicativa del Mapa Geologica del Peru," *Boletin*, Sociedad Geologica del Peru, Vol. 31, 88 pp.

Caelles, J. C., et al., 1971, "K-Ar Ages of Porphyry Copper Deposits and Associated Rocks in the Farllon Negro-Capillitas District, Catamarca, Argentina," *Economic Geology*, Vol. 66, pp. 961-964.

Clague, D. A., and Jarrard, R. D., 1973, "Tertiary Pacific Plate Motion Deduced from Hawaiian-Emperor Chain," *Bulletin*, Geological Society of America, Vol. 84, p. 1135.

Clark, A. H., 1976, Private Communication.

Clark, A. H., et al., 1970, "K-Ar Chronology of Granite Emplacement and Associated Mineralization, Copiapo Mining District, Atacama, Chile (Abstract)," *Economic Geology*, Vol. 65, p. 736.

Clark, A. H., and Farrar, E., 1973, "The Bolivian Tin Province," *Economic Geology*, Vol. 68, pp. 102-106.

Clark, A. H., and Zentilli, M., 1972, Annual Meeting Abstracts, Canadian Institute of Mining & Metallurgy, pp. 16-17.

Damon, J. E., 1973, Private Communication.

Dickinson, W. R., 1970, "Relations of Andesites, Granites and Derived Sandstones to Arc-Trench Tectonics," *Review of Geophysics and Space Physics*, Vol. 8, No. 4, pp. 813-850.

Ericksen, G. E., 1975, "Metallogenic Provinces of the Southeastern Pacific Region," Report IRCp-1, Open File, US Geological Survey.

Eyzaguirre, V. R., et al., 1975, "Age of Igneous Activity and Mineralization, Morococha District, Peru," *Economic Geology*, Vol. 70, p. 1123.

Farrar, E., et al., "K-Ar Evidence for the Post Paleozoic Migration of Granitic Intrusion Foci in the Andes of Northern Chile," *Earth and Planetary Science Letters*, Vol. 9, pp. 17-29.

Guilluly, J., 1971, "Plate Tectonics and Magmatic Evolution," *Bulletin*, Geological Society of America, Vol. 82, pp. 2386-2396.

Gonzales Bonorino, F., 1950, "Geologia y Petrografia de las Hojas 12d 6 13d, Provincia de Catamarca," *Boletin*, Director General, Industrial Minerals, Buenos Aires, Argentina.

Goossens, P. J., 1972, "Metallogeney in Ecuadorian Andes," *Economic Geology*, Vol. 67, p. 462.

Goossens, P. J., 1973, *Los Yacimientos e Indicios de les Minerales del Ecuador*, Universidad Guayaquil, Santiago de Guayaquil, Ecuador, p. 123.

Goossens, P. J., and Hollister V. F., 1973, "Chaucha," *Mineralium Deposita*, Vol. 8, No. 4.

Guest, J. E., 1969, "Upper Tertiary Ignimbrites in the Andean Cordillera, Chile," *Bulletin*, Geological Society of America, Vol. 80, pp. 337-362.

Gustafson, L. B., and Hunt, J. P., 1971, "Evolution of Mineralization at El Salvador, Chile (Abstract)," *Economic Geology*, Vol. 66, pp. 1266-1267.

Gustafson, L. B., and Hunt, J. P., 1975, "Porphyry Copper Deposit at El Salvador, Chile," *Economic Geology*, Vol. 70, pp. 857-912.

Haapala, P., 1949, "On Morochocha Breccias," Sociedad Geological del Peru, Vol. Jubilar 25, An. P11, pp. 2-11.

Hamilton, W. V., 1969, "The Volcanic Central Andes," *Proceedings of the Andesite Conference*, A.R. McBirney, ed., Oregon Dept. of Geology and Mineral Industries, Bulletin 65.

Haynes, S. J., 1972, "Relationship of Granite Chemistry to Magmatic Hydrothermal Ore Deposits, Andean Mobile Belt of Chile," Ph.D. Thesis, Queens University, Kingston, Ont., Canada.

Herron, E. M., 1972, "Seafloor Spreading and the Cenozoic History of the East Central Pacific," *Bulletin*, Geological Society of America, Vol. 83.

Hollister, V. F., 1974, "Regional Characteristics of Porphyry Copper Deposits of South America," *Trans. SME-AIME*, Vol. 256, p. 45.

Hollister, V. F., and Sirvas, E., 1974, "The Michiquillay Porphyry Copper," *Mineralium Deposita*, Vol. 9, No. 4, p. 261.

Hollister, V. F., and Bernstein, M., 1975, "Copaquire, Chile: Its Geological Setting and Porphyry Copper Deposit," *Trans. SME-AIME*, Vol. 258, p. 160.

Howell, F. H., and Molloy, J. S., 1960, "Geology of the Braden Ore Body, Chile, South America," *Economic Geology*, Vol. 55, pp. 863-906.

Irwin, W. P., and Coleman, R. G., 1972, Map No. MF 340, US Geological Survey.

James, D. E., 1971, "Plate Tectonic Model for the Evolution of the Central Andes," *Bulletin*, Geological Society of America, Vol. 82, pp. 3325-3346.

Knobler, R., and Werner, J., 1962, "The Mantos Blancos Operation," *Mining Engineering*, Vol. 14, No. 1, pp. 40-45.

Koukharsky, M., and Mirre, J. C., 1976, "Mi Vida," *Economic Geology*, Vol. 71, pp. 849-862.

Lacy, W. C., 1957, "Differentiation of Igneous Rocks and Ore Deposition in Peru," *Trans. AIME*, Vol. 208, p. 559.

Laughlin, A. W., Damon, P. E., and Watson, B. N., 1968, "K-Ar Dates from Toquepala and Michiquillay, Peru," *Economic Geology*, Vol. 63, pp. 166-168.

Le Pichon, X., and Hayes, D. E., 1971, "Marginal Offsets, Fractures, and Earl Opening of the South Atlantic," *Journal of Geophysical Research*, Vol. 76, No. 26, p. 487.

Lopez, V. M., 1942, "The Primary Mineralization at Chuquicamata, Chile," *Ore Deposits As Related to Structural Features*, W. H. Newhouse, ed., Princeton University Press, Princeton, NJ, pp. 126-128.

Lowell, J. D., 1974, "Three New Porphyry Copper Mines for Chile," *Mining Engineering,* Vol. 26, No. 11, p. 22.

Lowell, J. D., and Guilbert, J. M., 1970, "Lateral and Vertical Alteration-Mineralization Zoning in Porphyry Ore Deposits," *Economic Geology,* Vol. 65, pp. 373-408.

Ljungren, P., 1970, *Plan Cordillerano,* UNDP, Buenos Aires, Argentina.

Magliola-Mundet, H., 1964, "Le Gisement de Cuivre de Los Bronces de Disputada, Chile," *Chronique des Mines et de la Recherche Miniere,* Vol. 32, No. 330, pp. 120-127.

Maranzana, F., 1972, "Los Pelambres Hydrothermal Alteration Area, Chile," *Transactions,* Institution of Mining and Metallurgy, Feb., pp. B26-B33.

McCreary, E., 1970, "Rio Blanco," *Engineering and Mining Journal,* Dec., p. 94.

McCreary, E., 1971, "Cuajone Project Forges Ahead in Southern Peru," *Engineering and Mining Journal,* Dec., p. 72.

Muller-Kahle, E., and Damon, P. E., 1970, "K-Ar Age of Biotite Granodiorite Associated with Primary Cu-Mo Mineralization at Chaucha, Ecuador," *Correlation and Chronology of Ore Deposits and Volcanic Rocks,* Annotated Report C00-689-130, P. E. Damon, June, pp. 46-48.

Nagell, R. H., 1960, "Ore Controls in the Morococha District, Peru," *Economic Geology,* Vol. 55, pp. 962-984.

Navarro, H., 1968, Private Reports on MiVida (Costa Rico) Catamarca, Argentina.

Ocola, L. C., and Meyer, R. P., 1973, "Crustal Structure from the Pacific Basin to the Brazilian Shield Between 12° and 30° South," *Bulletin,* Geological Society of America, Vol. 84, pp. 3387-3404.

Palacios, C., and Oyarzun, R., 1975, "Relationship Between Depth of Benioff Zone and K and Sr Concentrations in Volcanic Rocks of Chile," *Geology,* Oct., p. 595.

Perry, V. D., 1952, "Geology of the Chuquicamata Ore Body," *Mining Engineering,* Vol. 4, No. 12, pp. 1166-1168.

Petersen, U., 1958, "Structure and Uplift of the Andes of Peru, Bolivia, Chile and Adjacent Parts of Argentina," *Boletin,* Sociedad Geologica del Peru, Vol. 33, pp. 57-129.

Petersen, U., 1965, "Regional Geology and Major Ore Deposits of Central Peru," *Economic Geology,* Vol. 60, pp. 407-476.

Richard, K. E., and Courtright, J. H., 1958, "Geology of Toquepala, Peru," *Trans. SME-AIME,* Vol. 211, pp. 262-266.

Rose, A. W., 1970, "Zonal Relations of Wallrock Alteration at Porphyry Copper Deposits,' *Economic Geology,* Vol. 65, p. 920.

Ridge, J. D., 1972, "Annotated Bibliographies of Mineral Deposits in the Western Hemisphere," *Memoir No. 131,* Geological Society of America.

Ruiz, F. C., et al., 1961, "Ages of Batholithic Intrusions of Northern and Central Chile," *Bulletin,* Geological Society of America, Vol. 72, pp. 1551-1560.

Ruiz, F. C., 1965, *Geologia y Yacimientos Metaliferos de Chile,* Inst. de Investigaciones Geologicas, Santiago, Chile.

Sales, R. H., and Meyer, C., 1948, "Wallrock Alteration at Butte, Montana," *Trans. AIME,* Vol. 178, pp. 9-35.

Segerstrom, K., 1967, "Geology and Ore Deposits of Central Atacama Province, Chile," *Bulletin,* Geological Society of America, Vol. 78, p. 305.

Sillitoe, R. H., and Sawkins, F. J., 1971, "Geologic, Mineralogic and Fluid Inclusion Studies Relating to the Origin of Copper-Bearing Tourmaline Breccia Pipes, Chile," *Economic Geology,* Vol. 66, pp. 1028-1041.

Sillitoe, R. H., 1972, "A Plate Tectonic Model for the Origin of Porphyry Copper Deposits," *Economic Geology,* Vol. 67, pp. 184-197.

Sillitoe, R. H., 1973, "Geology of the Los Pelambres Porphyry Copper Deposits, Chile," *Economic Geology,* Vol. 68, p. 1.

Terrones, L. A. J., 1958, "Structural Control of Contract Metasomatic Deposits in the Peruvian Cordillera," *Trans. SME-AIME,* Vol. 211, pp. 365-372.

Woollard, G. P., 1973, "Geological and Geophysical Setting of the Nazca Plate," *Abstracts,* Geological Society of America, p. 123.

YMAD (Yacimientos Mineros de Agua de Dionisio), 1969, *Concurso Publico de Propuestas-Bases,* Buenos Aires, Argentina, 39 pp.

Zentilli, M., 1974, "Geological Evolution and Metallogenic Relationships in the Andes of Northern Chile Between 26° and 29° South," Ph.D. Thesis, Queens University, Kingston, Ont., Canada.

2

Porphyry Copper Deposits of the Appalachian Orogen

CONTENTS

INTRODUCTION

Conditions for formation of porphyry copper deposits appear to have been propitious in the Appalachian orogen from the end of Precambrian into Middle Ordovician, and again from Middle Devonian into Mississippian. In the Appalachians we can now record the evolution of a porphyry copper province within a developing orogen, one that reached metamorphic and structural maturity by the end of Devonian. Fig. 15, after Irwin and Coleman (1972), locates that portion of the Appalachian orogen included in this study, and Fig. 16 illustrates general geology of the area. Deposits mentioned in Table 3 and in the text are shown in Fig. 16.

Data presented summarize personal observations by the author, published descriptions of specific areas noted in the bibliography, and abstracts from Provincial assessment files. Sources of previously published data are acknowledged. Few comprehensive descriptions of individual porphyry copper deposits in this area exist, so much of the data presented is new to the literature.

Porphyry mineralization has, for this chapter, been restricted to porphyry copper deposits, although porphyry molybdenum deposits (Clark, 1972) are also shown in Fig.

16 (Hollister, et al., 1974). Copper-molybdenum and tungsten-molybdenum skarns are also omitted to simplify discussion, though each is known in the area studied. Gaspe Copper is included, however; mineralization in this deposit is present in the intrusion as well as in the skarn adjacent (Bell and Scott, 1954).

Porphyry deposits listed in Table 3 are included because of some important feature, although other deposits have been reported (e.g., Kirkham and Soregaroli, 1975).

Clearly, no table now compiled could be complete. The pace of exploration insures that new occurrences will probably be found. Summaries of features exhibited by intrusions shown in Table 3 should remain valid, however, since they appear representative and are based on an adequate number of deposits in the area studied.

Paleozoic porphyry copper prospects (shown in Fig. 16) are more common, apparently, in the northeastern part of the orogen, rarely having been found south of Catheart or west of the Logan Line. Occurrences of molybdenum and other associated lithophile elements are, however, known to the south and west as well as further north. Killeen and Newman (1965), King (1970),

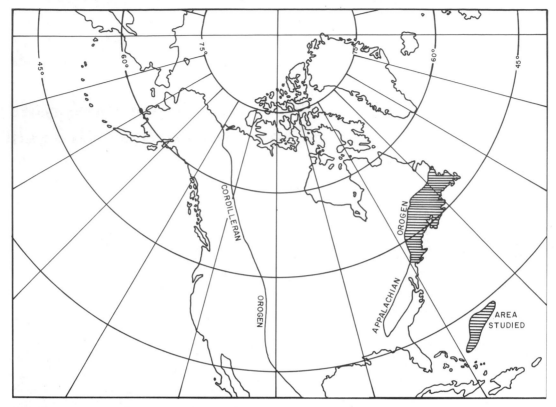

Fig. 15. Index map. The area studied in this chapter is shaded, but both the Cordilleran and Appalachian orogens are shown because the text refers to them.

and Kerr (1964) list lithophile metal occurrences south of the principal area of porphyry copper deposits. Although few occurrences listed by them appear to have a porphyry-deposit association, all show a strong affinity for magmatism in the Appalachian orogen.

The combination of lower grade hypogene mineralization, such as that found in deeper reaches of porphyry copper deposits like those in the Appalachians, and unfavorable conditions for widespread supergene sulfide accumulations in this area have left most known occurrences in the Appalachian orogen economically marginal at best. Thus, most deposits explored to date tend to be commercially unattractive under present conditions and ancillary data are not available, in contrast to the developed porphyry copper deposits of the Cordilleran region. Sulfur, oxygen, and Rb-Sr (rubidium-strontium) isotope analyses are skimpy. Conclusions reached here are not, therefore, abundantly

supported by detailed published descriptions and laboratory work.

Differences between alteration and structural characteristics described in the Lowell and Guilbert (1970) model and those known for the Appalachians may be ascribed to a greater age and greater erosion of the latter. Appalachian occurrences are labeled porphyry copper deposits because of the large mineralized areas involved, the typical pervasive potassium metasomatism, the chalcopyrite-molybdenite mineralogy, and the basic petrographic similarity of these intrusions to porphyries of other regions. The Lowell and Guilbert (1970) model probably should be expanded to accommodate characteristics of the Appalachian deposits.

Terms

To clarify geological terms, several definitions are cited from the literature. The Logan Line is explained by Fortier and Hogson

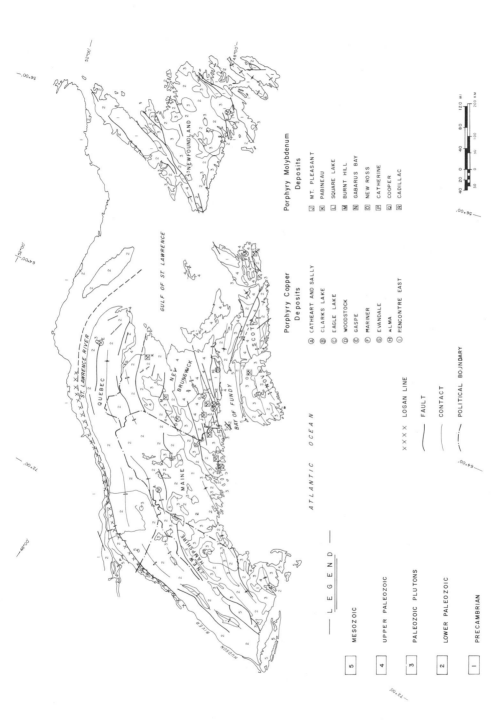

Fig. 16. Generalized geology of the northeastern Appalachians. Porphyry copper deposits discussed in the text are shown. Porphyry molybdenum deposits are included for those who wish to refer to them. Simplification of the geology into five units emphasizes the plutons (*after numerous US Geological Survey, Geological Society of Canada, state and provincial government surveys*).

Table 3. Porphyry Copper Occurrences in the Appalachians

Name	Location	Province or State	Pluton Type	Age	Host	Alteration Zone from Core	Pyrite Zone Size	Content, %	Structure
Sally Mountain	45°35', 70°20'	Maine	Qtz Por	Ordov	Camb	Pot Prop	None		Stockwork
Catheart	45°35', 70°10'	Maine	Qtz Por	457 m.y.	Camb	Pot Phy Prop	1.0x1.5 km	3	Stockwork
Clarks Lake	45°17', 66°13'	New Bruns	Qtz Mon Por	Dev	Precamb	Pot Phy Prop	1.5x1.5 km	3	Stockwork
Woodstock	46°00', 67°35'	New Bruns	Qtz Mon Por	Dev	Ordov	Pot Phy Prop	1.0x1.0 km	3	Stockwork
Gaspe	48°58', 65°30'	Quebec	Qtz Mon	350 m.y.	Dev	Pot Phy Prop	Erratic	5	Stockwork
Evandale	45°35', 66°04'	New Bruns	Qtz Mon	364 m.y.	Ordov	Pot Prop	None		Stockwork
Eagle Lake	45°15', 66°22'	New Bruns	Qtz Mon	Dev	Precamb	Pot Phy Prop	Erratic		Stockwork
Mariner	46°05', 60°20'	Nova Scotia	Qtz Mon	584 m.y.	Camb	Pot Prop	None		Stockwork
Alma	45°35', 64°56'	New Bruns	Qtz Mon	Camb(?)	Precamb	Pot Prop	None		Stockwork
Rencontre East	47°30', 55°10'	Newfound	Qtz Mon Por	Dev	Dev	Pot Phy Prop	1.0x2.0 km	4	Stockwork

Bruns—Brunswick
Newfound—Newfoundland
Qtz—Quartz
Mon—Monzonite
Por—Porphyry

Ordov—Ordovician
m.y.—million years
Dev—Devonian
Camb—Cambrian
Precamb—Precambrian

Prop—Propylitic
Pot—Potassic
Phy—Phyllic
km—kilometer

(1972, p. 11) as the southeastern limit of the Precambrian Shield. For the northern Appalachians and most of the area of this study, Howie and Cumming (1963) place the Appalachian orogen between the Logan Line and the Atlantic. Southward from the Catheart deposit, confinement of the Appalachian orogen to the area between the Logan Line and the Atlantic Ocean is less definitive. To the north, however, this feature separates Grenville subprovince rocks from the Appalachian orogen. Dewey (1969) defines the Appalachian orogen itself to include various episodes of Paleozoic tectonism found within the area studied; for this chapter, the region considered extends from the Hudson River north to the eastern coast of Newfoundland. For purposes of this study, the Taconic orogeny culminated at about 450 m.y. and the Acadian extended from about 394 to 360 m.y. (Williams, et al., 1972, and Pajari, et al., 1974). Fig. 16 shows the distribution of plutons related to these orogenies, but also includes other plutons whose age may not be synchronous with an orogenic event.

Porphyry copper deposits are large bodies of mineralized rock, usually including a porphyritic intrusive phase, that have copper sulfide minerals disseminated through them, according to Lowell and Guilbert (1970). Molybdenum nearly always accompanies copper in Appalachian deposits in the Lowell and Guilbert (1970) model, and may be economically dominant in value locally. The Lowell and Guilbert (1970) model may be expanded to include mineralized plutons with the texture and alteration mineralogy of deposits of the Appalachian orogen. Porphyry-type mineralization is the characteristic dissemination of metallic minerals in a porphyry setting, with metallic sulfides occurring as true disseminations, on fractures, or in veins, veinlets, or breccias. Table 3 classifies stockwork and breccia pipe as primary structural types after Sillitoe and Sawkins (1971). Porphyry deposits may have metallic sulfides either in an intersecting network of veinlets or fracture surfaces (stockwork type), or predominantly as matrix filling or fracture coatings in a brecciated rock mass

(breccia type). True disseminated mineralization is important in Appalachian deposits. In addition, the word ore is used in a strictly noneconomic sense to encompass the spectrum of mineralization associated with these deposits, exclusive of past, present, or future potential for exploitation.

GEOLOGIC SETTING

Regional geology of the Appalachians as it concerns porphyry copper deposits is briefly summarized. The northern Appalachian orogen is bounded on the west by the Grenville subprovince of the Canadian Shield, whose metamorphism Wanless (1969) shows to be 1000 m.y. (Helikian) or older. The east appears bounded by rocks that Williams, et al. (1972), show to be about 600 m.y. (Hadrynian). Between these two segments of the Precambrian, formations from Cambrian to Middle Ordovician age were involved in Taconic orogeny, and—together with younger rocks—in the Middle Devonian Acadian orogeny. Mississippian to Mesozoic rocks were deformed locally in post-Acadian disturbances. Complete descriptions of the orogen are given in Zen, et al. (1968), or Rodgers (1970).

Part of the orogen in which porphyry deposits formed lacks some characteristics attributed to plate-consuming subduction zones. Large intrusions of the Sierra Nevada batholithic type are scarce, according to Zietz and Zen (1973, p. 27). Cady (1972, p. 3808) also notes paucity of andesitic volcanics "commonly associated with island arc systems adjacent to subduction zones" in the Paleozoic section. By analogy, major porphyry copper deposits of the Cordilleran orogen in the US also are not associated with batholitic-scale intrusions or island arc volcanic environments, but tend to be spatially separate from these subduction zone manifestations. Furthermore, Zietz and Zen (1973) apparently view the portion of crust penetrated by porphyry copper-associated magmas as a thick section of continental crust dominated by metasediments and metasilicic volcanics, with rare oceanic volcanics. From the Logan Line east, except for Newfoundland and parts of Quebec, the orogen is essentially ensialic; and most porphyry copper deposits occur in this area.

Cady (1972) convincingly argues for "distensional tectonic conditions" for much of the Appalachian orogen during the Cambrian and Lower and Middle Ordovician. Coincidentally, interpretation of radiometric dating and stratigraphic data suggests that at least the Mariner, Sally Mountain, and Catheart porphyries formed during this interval. The latter two appeared near the culmination of the Taconic orogeny and are associated with Oliverian and Highlandcroft igneous events (Nayler, 1968, and Boone, et al., 1970). Woodstock and Gaspe appear to be dated just after culmination of the Acadian orogeny, again under distensional conditions (Fig. 23).

Burchfiel and Davis (1972) imply distensional tectonic conditions at the time when and location where most major porphyry copper mineralization occurs in the Cordilleran orogen of the US. Extensional features found in Cordilleran porphyry copper deposits compiled by Lowell and Guilbert (1970) reflect this tectonic environment. Intramineral transcurrent fault motion during porphyry mineralization documented by Morris (1976) also suggests that similar features are found in Appalachian orogen occurrences.

The Taconic and Acadian orogenies, two brief episodes of compressive tectonics, took place within the 580-320 m.y. period when porphyry-type deposits developed in the Appalachian orogen. Each was accompanied by folding, crustal shortening, and an inferred thickening of the continental crust. It may be significant that no known porphyry molybdenum deposits precede a 400 m.y. date, as do porphyry copper deposits. Perhaps thickening of the continental crust was essential to formation of porphyry molybdenum deposits and hence none formed prior to the Taconic orogeny. Molybdenum-bearing porphyry copper deposits, on the other hand, date from 580 m.y. (Mariner) to 350 m.y. (Gaspe), so deposits apparently formed prior to as well as after periods of crustal thickening. Sufficient deposits are now dated to make significant the relationship between crustal evolution and type of porphyry mineralization.

The Mount Pleasant porphyry molybdenum occurrence, with a 320 m.y. K-Ar date, is Lower Mississippian on stratigraphic evidence (Ruitenburg, 1972). Thus porphyry-type mineralization persisted into the upper Paleozoic in the northern Appalachian orogen.

Wynne-Edwards and Hasan (1970) and Cady (1972) present evidence relating Paleozoic orogenies of the Appalachians with the Caledonian and Hercynian in Europe. Porphyry-type molybdenum-lithophile deposits exist in the Hercynian of Portugal, as detailed in the Cotelo Neiva (1972) description, and in Spain, according to Sierra, et al. (1972). It is beyond the scope of this chapter, however, to consider the Caledonian or Hercynian or the porphyry copper deposits associated with those orogenies in Europe.

Williams, et al. (1972) minimized the names Taconic and Acadian in grouping igneous rocks in the Appalachian orogen. They used Precambrian, late Precambrian to early Paleozoic, and Silurian-Devonian. Their grouping for intrusions appears more flexible than a grouping based on orogenies. In any case, porphyry-type mineralization and the intermediate to acid plutons associated with sulfide concentrations appear entirely Paleozoic. They are part of the second phase of Potter's (1970) Appalachian metallogeny.

Descriptions of Individual Deposits

General: The following descriptions of the Mariner, Gaspe, Evandale, Eagle Lake, Clarks Lake, and Sally Mountain deposits provide a basis for generalization on all Appalachian porphyry coppers. In addition, certain features of other deposits are briefly mentioned where these are well known and particularly significant. Economic considerations were not paramount in the examples chosen. A description is also given for mineralized intrusions at Buchans, because Kirkham and Soregaroli (1975) have suggested this may be a porphyry deposit.

Mariner Mines, Nova Scotia: The Mariner mines' porphyry copper deposit in Nova

Scotia occurs in a zoned diorite-quartz monzonite stock (the Coxheath pluton) that intrudes an apparently coeval, dominantly andesitic volcanic pile. Volcanics are associated with latest Precambrian to earliest Cambrian sediments (Helmstaedt and Tella, 1973).

From a dominantly dioritic outer margin, the stock grades through quartz diorite to a quartz monzonite core. Gradational contacts between various phases suggest that zoning may be the result of differentiation. Other workers have used diabase (as in Fig. 17) or

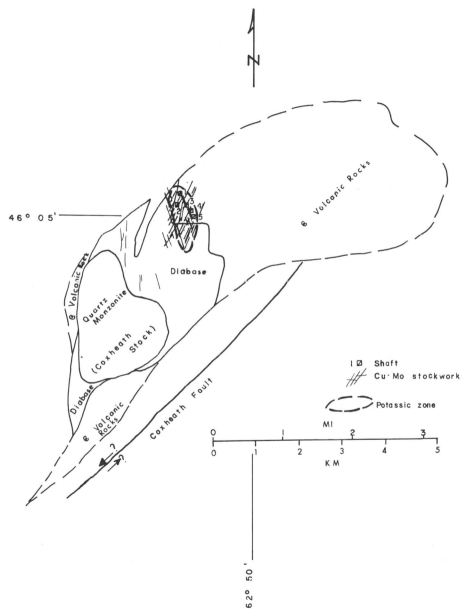

Fig. 17. Geologic outline of Mariner mines, N. S. The core and the outline of the pluton are shown with the mineralized area. The veinlet trend in and location of the stockwork are independent of the core phase but may be penecontemporaneous with it. Shafts are numbered and the rock called diabase contains diorite and quartz diorite phases of pluton (*from the Nova Scotia Dept. of Mines Assessment Files*).

quartz gabbro for the outer zone and granite for the inner; however, petrographic studies (Hollister, et al., 1974) find these terms incorrect. The quartz monzonite core of the stock is essentially fresh and free of hydrothermal alteration. A potassic zone measuring 608 x 304 m (2000 x 1000 ft) lies north of the quartz monzonite core and contains the porphyry copper mineralization. Rock alteration minerals are orthoclase and epidote, with lesser chlorite. Mineralization in the zone is confined to closely spaced quartz-orthoclase-epidote-sulfide veinlets, most of which strike N10E. Other veinlet orientations are present, and postmineral regional compression has folded some of the N10E veinlets, but these variations comprise less than 10% of the total number of mineralized fractures. Copper-bearing veins are typically composed of quartz, epidote, orthoclase, chalcopyrite, bornite, and rarely pyrite. Chalcopyrite and sparse pyrite are disseminated in the wall rock. Copper and molybdenum sulfides tend to be separate, with molybdenum-bearing veins typically composed of quartz, orthoclase, and molybdenite. Pyrite is not conspicuous in these fractures.

A K-Ar age of 584 m.y. on Geological Survey of Canada sample 66-160 from the mineralized area (Wanless, et al., 1968) agrees with an Rb-Sr date of 580 m.y. on a whole rock-feldspar pair from the quartz monzonite (Cormier, 1972), suggesting that both intrusion and mineralization took place during very Early Cambrian or late Precambrian.

Suggestion of magmatic differentiation inferred in the zoned stock raises the possibility that mineralization also could have been an end product of this process. Field relationships suggest that the quartz monzonite intrusion and hydrothermal events were the last to occur in the volcanic-plutonic-hydrothermal association, and that these two were probably contemporaneous, with no firm evidence to uniquely link magmatic differentiation and hydrothermal activity to Mariner.

Enveloping the potassic zone, a fairly typical propylitic zone contains calcite-chlorite-filled fractures that also strike N10E. No phyllic or argillic zones exist at Mariner. The geologic setting is shown in Fig. 17.

The coarse-grained, although usually equigranular, nature of the alteration minerals, the presence of a large potassic zone (but the absence of sericite), the appearance of ore sulfides as disseminations in the altered zone, and the existence of a fresh, unaltered core of stock adjacent to the potassic zone combine to suggest that present exposures at Mariner represent the roots of the ore deposit.

Sally Mountain, Maine: Oliverian and Highlandcroft intrusions and felsic metavolcanics in northern New England have been described by Naylor (1968) and Boone, et al. (1970). The Catheart and Sally Mountain deposits are associated with Attean quartz monzonite in this group. The Attean intrusion is a leucocratic, coarse-grained, and locally deformed quartz monzonite (Albee and Boudette, 1972), with a Pb-U date of 470 m.y. (Boone, et al., 1970, p. 19), or Ordovician age. Silurian conglomerate lies on quartz monzonite on the north flank of Sally Mountain; thus an Ordovician age seems reasonable.

Dikes and irregular bodies of quartz porphyry, granophyre, and fine-grained quartz monzonite intrude the Attean quartz monzonite at Catheart and Sally Mountain. Porphyritic textures of quartz porphyries suggest an epizonal emplacement. Progressive enrichment in silica and alkalies from the oldest phase (Attean quartz monzonite) to the youngest (quartz porphyry) is compatible with derivation of quartz porphyry from quartz monzonite by differentiation. Schmidt (1974) does not admit such a genesis, although he does not rule it out.

The quartz porphyry plug at Sally Mountain (Fig. 18) appears to have forcefully penetrated the Attean quartz monzonite and produced a radiating set of fractures upon which a strongly developed N20E-trending set has been superimposed. The Sally Mountain fault appears to have a regional control because all younger intrusions appear elongated along it.

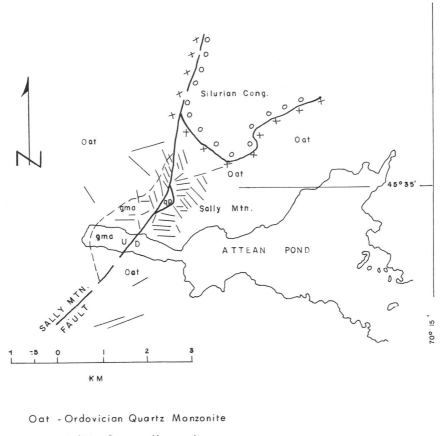

Oat - Ordovician Quartz Monzonite

gma - Aplitic Quartz Monzonite

qp - Quartz Porphyry

- Veins

Fig. 18. Structural elements of Sally Mountain, ME. Mineralized fractures commonly radiate from the central quartz porphyry plug. Potassic alteration appears pervasively well developed in the walls of the fractures but sulfide mineralization is largely confined to the fractures themselves. The present exposures are believed to represent the roots of a porphyry copper system because no phyllic or argillic zones accompany the potassic zone and because of the impoverishment of disseminated sulfide (*modified from various US Geological Survey sources*).

Chalcopyrite and molybdenite mineralization is largely confined to fractures that cut the quartz porphyry and quartz monzonite. Both rock types are nearly free of sulfide dissemination, with only minor molybdenite dissemination found in the quartz porphyry. Quartz, orthoclase, microcline, pyrite, chlorite, and minor epidote developed with the sulfides. Albee and Boudette (1972) describe microcline phenocrysts in the wall rocks. Pervasive pyritization is nearly absent, as are the usual products of a phyllic zone. Potassic zone minerals (including microcline and chlorite) are surrounded by a zone containing typical propylitic zone minerals (chlorite, calcite, zoisite). The large area containing sulfide-bearing fractures plus the presence of potassic zone silicates lining the fracture walls suggests that the present level of erosion may have exposed the roots of a porphyry copper deposit.

Catheart, Maine: The Catheart porphyry

copper occurs in and near a quartz porphyry plug that intrudes several phases of Attean quartz monzonite (Schmidt, 1974). Distinctions between the quartz monzonite phases are textural. The quartz porphyry intrusion most closely related, temporally and spatially, to ore contains about 3% disseminated sulfides, which are mostly pyrite but include sparse chalcopyrite.

Fracturing in and near the quartz porphyry is dominated by an east-west trending quartz-sulfide stockwork. Other veinlet trends, most notably a N20E set, are of secondary importance. Coincident with the core of the stockwork, the Attean complex is altered; secondary K-feldspar, biotite, and chlorite are developed in a potassic zone assemblage. Microcline is prominent. Adjacent to and surrounding the potassic zone is a phyllic zone that contains about 4% pyrite, sericite (with both fine and coarse crystals), and quartz. Outside the phyllic zone and gradational with it, a propylitic zone is marked by development of chlorite, kaolin, calcite, and epidote. No argillic zone has been detected as such.

Mineralization in the potassic zone, based on 116 drill holes in and near the ore body, is chalcopyrite and molybdenite in fracture fillings and disseminations. Grade of the potassic zone is about 0.2% Cu plus minor molybdenum. The interface between the potassic and phyllic zones contains the best copper and molybdenum values, with an average of 0.3% Cu and 0.06% Mo in a persistent but thin shell around the potassic zone. Stannite occurs erratically but weakly in veins in the phyllic zone. Outward from this envelope, grades of both chalcopyrite and molybdenite decrease rapidly. Sphalerite occurs east of the copper, in the propylitic zone, suggesting rudimentary metal zoning. Erratic supergene chalcocite mineralization occurs in isolated sections near surface portions of the deposit.

Eagle Lake, New Brunswick: The Eagle Lake deposit, according to Ruitenberg (1969) and Potter (1973), is situated in a biotite quartz monzonite that cuts Precambrian(?) metavolcanics. The stock is elongated subparallel to the Beaver Harbour-Long

Reach fault. One set of fractures trends northeast, paralleling the fault, and a more extensive set has an average N5W trend. Strongest mineralization occurs on the northeast trending set, though only in the area of intersection of the two sets. Mineralization consists of chalcopyrite and molybdenite in veinlets with rare galena in a gangue of quartz, orthoclase, sericite, and pyrite.

Alteration around individual mineralized fractures includes pervasive orthoclase and chlorite (i.e., potassic zone) replacements of quartz monzonite, giving it either a pale greenish or a deep pink color, or, more erratically, a patchy quartz-sericite (or phyllic zone) alteration assemblage. Quartz-sericite products are restricted to the area of intersection of fractures, whereas both zones are surrounded by a chlorite-dominant propylitic zone. General geologic relationships are shown in Fig. 19. Hypogene sulfide occurs at or close to the surface due to glacial removal of any preglacial leached capping.

Clarks Lake, New Brunswick: The porphyry copper deposit at Clarks Lake occurs in a composite intrusion of porphyritic diorite, quartz diorite, and quartz monzonite with quartz porphyry. Exposures are poor, and it is not possible to determine the nature of contacts or the limits of any alteration zone in the complex. The quartz porphyry may be an altered leucocratic quartz monzonite porphyry.

Mineralization appears centered about and within quartz monzonite porphyry and consists primarily of quartz and sulfide filled fractures, which may also carry orthoclase or sericite. Sparse wall rock dissemination of chalcopyrite is visible in outcrop. Dominant veinlet trend is N5E; N50W and N60E sets also are identifiable. These features are shown in Fig. 20.

Alteration suites consist of an orthoclase-chlorite pervasive replacement (potassic zone), an adjacent quartz-sericite-pyrite alteration area (phyllic zone), and a peripheral propylitic zone. Pyrite content in the phyllic zone averages about 3%. Sulfide occurs at or close to the surface and any leached

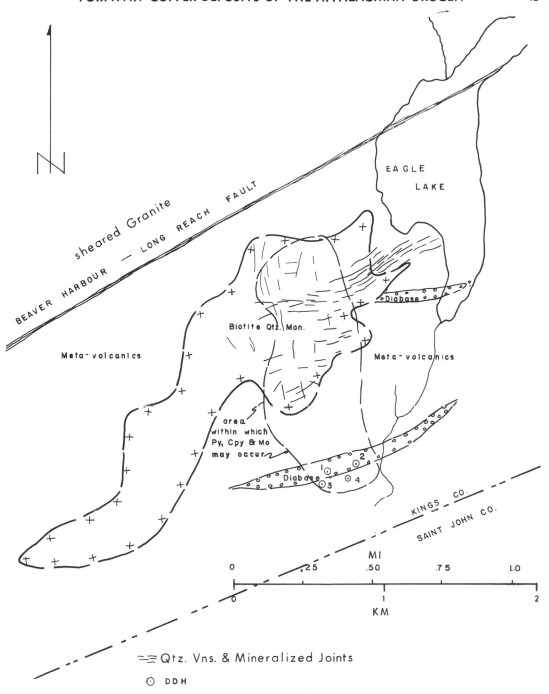

Fig. 19. General geology of Eagle Lake, N.B. A biotite quartz monzonite stock intrudes Precambrian(?) metavolcanics south of the subparallel to the Beaver Harbour-Long Reach fault. Mineralized fractures widely dispersed in areas of potassic alteration occur within this intrusion and satellitic diabase.

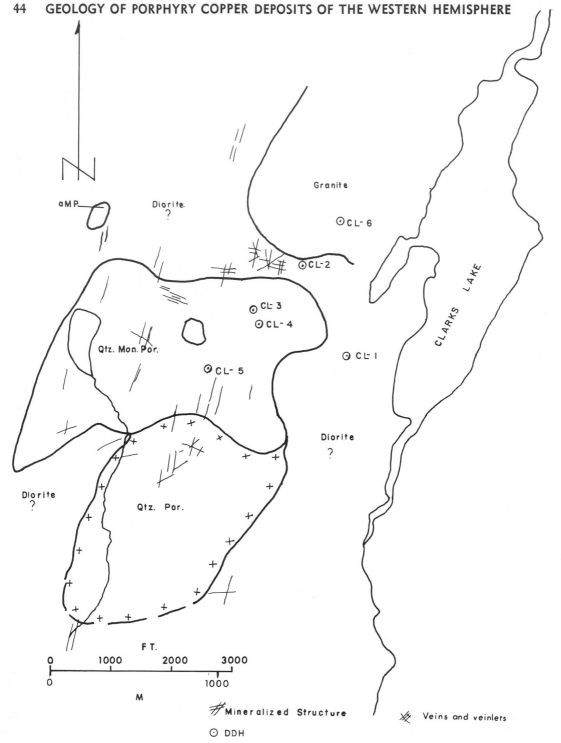

Fig. 20. General geology of Clarks Lake, N.B. A composite quartz diorite-granodiorite-quartz monzonite-quartz porphyry intrusive center outcrops poorly near Clarks Lake. Porphyry-type phyllic and potassic alteration zones are cut by mineralized fractures and stockwork.

capping that may have existed has been removed by glaciation.

Seven holes have been drilled in the potassic and phyllic alteration zones. These show subcommercial copper and molybdenum values in both zones. Deepest penetration was in phyllic rock that was pyritized but showed no copper or molybdenum grade improvement with depth.

The widespread but weak level of pyrite dissemination, when viewed in the perspective of a large area of potassic alteration with a low, but fairly uniform level of metallization relative to a small phyllic zone surrounding a fresh, unaltered intrusive center, suggests that this deposit is at the base of a large eroded porphyry system. Byproducts of mineralization (silicate alteration and pyrite) are significantly present, whereas the sulfide ore minerals themselves appear to have been deposited elsewhere. Drilling suggests further weakening of total sulfide with depth.

Evandale, New Brunswick: The porphyry copper deposit at Evandale is associated with a nearly circular quartz monzonite piercement into Ordovician argillite with a K-Ar age of 364 m.y. (Ruitenberg, 1972).

The piercement or diapir is cut by the Hampstead shear zone, a N25W-trending regional fault. Copper and molybdenum sulfide may occur in and are confined to arcuate ring fractures near the margin of the stock or in fault fracture filling in the Hampstead shear zone. Most prominent mineralization appears at the intersection of the Hampstead shear and marginal arcuate fractures of the stock. Chalcopyrite and molybdenite occur infrequently in quartz and quartz-pyrite veinlets, and minor chalcopyrite is disseminated in quartz monzonite near the fractures. Fig. 21 shows general distribution of veinlets as well as the potassic zone.

Alteration zoning in the Evandale stock is particularly well developed. Core of the intrusion is a fresh unaltered normal quartz monzonite with a typical orthoclase to plagioclase ratio. As the margin is approached, however, orthoclase content increases gradually into the area with the field designation

potash rich quartz monzonite (see Fig. 21). This is essentially a mixture of orthoclase, quartz, and biotite, with plagioclase decreasing as orthoclase increases. Parts of this zone contain chlorite and epidote in place of biotite. The potassic alteration zone encompasses all mineralized fractures, completely surrounds the fresh core, and extends to where the stock makes contact with argillite. Argillite appears to constitute the typical propylitic zone found in other porphyry copper centers, because the alteration zone mineralogy associated with mineralized fractures is dominantly chloritic. Pyrite is not a significant disseminated mineral, and phyllic and argillic zones of the Lowell and Guilbert (1970) model are missing.

The very large visible potassic zone and the absence of a phyllic zone suggest that erosion has cut deeply into this deposit, removing the phyllic zone. Presence of a fresh unaltered core is compatible with such an assumption. Weak development of sulfide, though widespread in the potassic zone, also is to be expected in such a root zone. Paucity of sulfide in this system permits speculation that it is not part of a porphyry copper deposit. On the other hand, presence of a chalcopyrite-molybdenite couple favors porphyry affinity.

Gaspe Copper, Quebec: The Copper Mountain deposit at Gaspe Copper has been described by Bell and Scott (1954) and McAllister and Lamarche (1972). Fig. 22 is modified from McAllister and Lamarche (1972) to fit more recent mine developments.

In essence, the deposit consists of a stock and dikes of quartz monzonite in Lower Devonian silt, shale, and limestone. Minor diabase dikes also occur in the ore zone. The stock has been dated Devonian.

Although hypogene copper sulfide occurs in parts of the intrusions, all known ore occurs in sediments, primarily as a bedded replacement in highly altered calcareous siltstone and limestone, and as stockwork in altered siltstone. Presence of mineralization in the intrusion, with evident potassic and hydrogen metasomatism, qualify this deposit

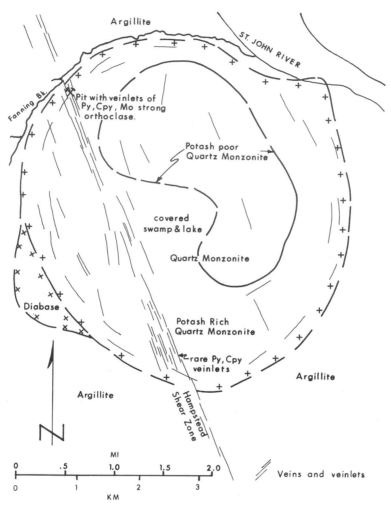

Fig. 21. Geologic outline of the Evandale stock, N.B. Porphyry copper-type mineralization occurs in outer margins of the Evandale stock where the Hampstead shear cuts the potassic (secondary orthoclase) zone. Veins and small stockworks carrying copper and molybdenum sulfides can be seen within the potassic-altered phase of the stock. The large potassic zone, the absence of both phyllic and argillic alteration, and the impoverishment in sulfide in the potassic zone are compatible with present exposures being considered the roots of the deposit (*modified from New Brunswick Dept. of Natural Resources Assessment Files*).

as a porphyry copper. At Copper Mountain, stockwork is the most important locus for ore. The principal trend of stockwork veinlets is N29W, with steep westerly dips. This trend is normal to fold axes in the region, suggesting a possible genetic tie between N29W-oriented compression and intrusion and mineralization. Confusing this picture is an east-west striking quartz-monzonite porphyry dike north of the main stock; the best values appear spatially associated with it (see Fig. 22). The area containing ore appears to average 608 x 304 m (2000 x 1000 ft), and in the hypogene zone contains about 5% pyrite with chalcopyrite and molybdenite. Molybdenum values improve with depth. Minor scheelite also has been identified in the ores. Most chalcopyrite occurs as fracture filling,

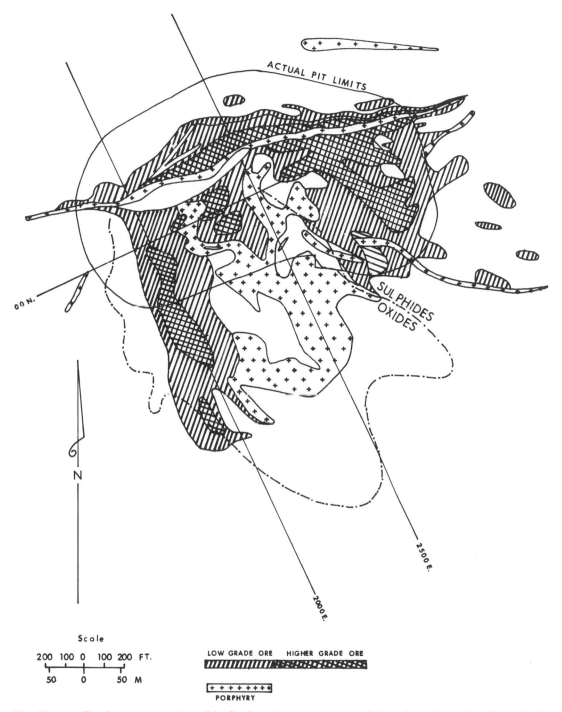

Fig. 22. Metallization at Gaspe, Que. Distribution of copper ore around plutonic rocks on the 597-m (1960) level indicates proximity of ore to igneous rocks. Oxidation may be partially Carboniferous as well as Tertiary (*modified after McAllister and LeMarch, 1972*).

although minor dissemination of both chalcopyrite and molybdenite occurs in the intrusions.

Alteration minerals appear to be a function of rock type, so application of the simple zoning pattern from the Lowell and Guilbert (1970) model is somewhat misleading. The most striking feature of the alteration halo is its intense silicification. Bell and Scott (1954) estimate that at least 453.5 million mt (500 million tons) of silica were added to the rocks during alteration.

Buchans, Newfoundland: Extensive Ordovician polymetallic volcanogenic ore deposits at Buchans surround an eroded mineralized volcanic center. Kirkham and Soregaroli (1975) note the similarity of some pyritic stockworks at Buchans to porphyry copper deposits. Affinities between pyritic stockworks in the volcanic center and porphyry copper deposits justify their mention, although such alteration and mineralization centers are common to most large volcanogenic massive sulfide deposits. To suggest that this example is a porphyry copper deposit would require reclassification of other deposits of cross-cutting massive sulfide mineralization as porphyry deposits. Difficulties in establishing a new porphyry model for such deposits makes acceptance of such a classification remote. For this reason, pyrite stockworks at Buchans are not included with Appalachian orogen porphyry copper deposits.

Porphyry Copper Deposits of the Southern Appalachians: Beyond the area included in this study, in the southern Appalachians of North and South Carolina, known porphyry copper deposits have sulfide mineralogy and alteration mineral assemblages similar to those described in the northeast (Beg, 1975, and Cook, 1971). The Catawaba deposit, drilled by Continental Oil, has an Rb-Sr date of 325 m.y. It contains a well developed potassic zone and a smaller phyllic zone in porphyritic granodiorite. The potassic zone contains orthoclase, chlorite, and minor epidote; and the phyllic zone has pyrite and characteristic sericite with prominent quartz (Beg, 1975). An outer epidote-chlorite pro-

pylitic zone encircles other assemblages.

The Conner deposit is associated with biotite quartz monzonite porphyry (Cook, 1971) and has epidote in a potassic zone dominated by chlorite orthoclase. Sericite and quartz abound in the adjacent phyllic zone, and some epidote is visible in this assemblàge. An epidote-bearing propylitic zone encloses other alteration assemblages.

The Newell occurrence described by Worthington (1973) consists of quartz monzonite stock intruding gneissic rock, with potassic, phyllic, and propylitic alteration developed largely in gneiss.

In each of these deposits, the potassic zone, large relative to the phyllic zone, and the appearance of epidote together with a very widespread but weakly disseminated pyrite halo are taken as evidence that exposed deposits represent roots of porphyry systems (Beg, 1975). Additionally, drilling results at Catawaba apparently show downward weakening of sulfide mineralization as well as deterioration of potassic metasomatism.

Characteristics of Appalachian Porphyry Copper Deposits

Petrography: Table 3 shows the majority of intrusions closely associated with Appalachian orogen porphyry copper-type mineralization to be quartz monzontic in composition. In most examples, intrusions display a fresh, unaltered, and unmineralized core located well within the altered and mineralized system. In composition they approach the average of the Lowell and Guilbert (1970) model. Occasionally, however, minor variations may be found, such as Deboullie, described by Boone (1962), or the quartz porphyries of Catheart and Sally Mountain. Except for porphyries at Catheart, Sally Mountain, Rencontre East, and Woodstock, the intrusions usually are not porphyritic but are equigranular hypidiomorphic in texture. Magmatic biotite occurs in all host rocks, and hornblende is a common mafic. Pyrite may accompany silicates as an apparent primary magmatic mineral. Also, rare disseminated chalcopyrite is an apparently magmatic constituent in igneous rocks that appear free of hydrothermal alteration. Orthoclase, plagioclase, and quartz all appear in variable amounts as primary constituents.

A clear connection between mineralized intrusions and an extrusive volcanic sequence approximately the same age cannot be established for most deposits. The Mariner stock is bounded by Cambrian early andesitic to late dacitic continental calcalkalic volcanics that may be genetically related. A similar relationship between volcanic rocks and a differentiated intrusion may exist for Sally Mountain and Catheart but is not evident for Woodstock, Eagle Lake, Clarks Lake, Evandale, or Gaspe. If volcanics accompanied any intrusions, these volcanics have been removed by erosion.

Alteration: As a group, Appalachian porphyry copper deposits have alteration zones generally compatible with the Lowell and Guilbert (1970) model and commonly adjacent to relatively unaltered core rocks of intrusive complexes. Alteration zones display some significant differences from characteristics predicted by the model. Postmineral metamorphism and other events during the long history since deposition may account for some differences. This section will briefly examine similarities and attempt to reconcile differences.

The potassic zone of Cordilleran orogen porphyry copper deposits is dominated by hydrothermal orthoclase and biotite, with subsidiary orthoclase-chlorite zones in some cases. In Appalachian porphyries, microcline may occur in place of all or part of the orthoclase. Original orthoclase may have inverted totally or partially to microcline. In the Catheart, Sally Mountain (Albee and Boudette, 1972), and Rencontre East (Snelgrove and Baird, 1953) porphyries, microcline forms as phenocrysts. Most intrusions that contain microcline or orthoclase also contain biotite, chlorite, or epidote-chlorite mixtures. In some intrusions (e.g., Mariner) the potassic zone is a pink mixture of orthoclase with minor epidote and chlorite. Sani-

dine has also been identified in the potassic zone at Rencontre East (White, 1940). Differences between potassic zones of the Appalachian and Cordilleran orogens appear to be mineralogic, not chemical. It appears, therefore, that in Appalachian deposits a K-feldspar-epidote (or epidote-chlorite) zone may grade outwardly into a K-feldspar-chlorite zone or a K-feldspar-biotite zone.

A few Appalachian deposits have identifiable phyllic zones. These usually occur as poorly defined incomplete concentric cylinders or cones around a potassic zone, while the potassic zone itself may either surround or occur on one side of a fresh core intrusion. Sericite in the phyllic zone is intimately intergrown with quartz and pyrite in Appalachian orogen deposits, and quartz-sericite veinlets are similar to those of Cordilleran porphyries. Two modes of occurrence of sericite are recognized: a fine-grained pervasive sericite (e.g., Woodstock and Clarks Lake), which is typical of phyllic zones of porphyry copper deposits of the Cordilleran orogen, and a coarse-grained sericite that lines veinlet walls in some deposits (e.g., Catheart).

Argillic zones are missing in Appalachian deposits. The propylitic zone of Appalachian deposits appears to have the same mineralogy as that of the Lowell and Guilbert (1970) model and is dominated by chlorite. Propylitic zones are peripheral to potassic zones or to phyllic zones if such zones exist.

The quantity of pyrite in the halos of Appalachian deposits generally is appreciably less than in those of the Cordilleran orogen. In rare occurrences, as at Rencontre East (Vokes, 1963), Woodstock, Gaspe, Clarks Lake, and Catheart, pyrite is strongly developed in pervasive disseminations. Pyrite is difficult to evaluate because it weathers easily, but for most Appalachian deposits it is less pronounced than would be predicted by the Lowell and Guilbert (1970) model.

In summary, alteration zones in Appalachian porphyry copper deposits generally may be described as successive cylinders or cones that incompletely surround or occur to one side of a fresh core intrusion. Adjacent to the fresh intrusion may be the K-feldspar-epidote or epidote-chlorite zone. Proceeding outward from this, a K-feldspar-chlorite zone and then a K-feldspar-biotite zone may occur. Any or all individual zones may be missing. Where present, a phyllic zone, usually small and low in pyrite, is external to these facies of the potassic zone. The argillic zone is absent and the propylitic zone is adjacent to either the phyllic or potassic zone.

Mineralization and Structure: In the vast majority of Appalachian porphyry copper deposits, metallic sulfide mineralization is partly disseminated as rock mineral and partly as filling in veins and veinlets, according to Ruitenberg (1969). These deposits may be broadly grouped in the stockwork model. In the narrowest sense they are rarely true stockworks but, rather, areas where multiple veining occurs. In actual distribution of hypogene copper and molybdenum sulfides, true disseminations may account for more than 50% of the values in the potassic zone of some porphyry deposits. In most Appalachian orogen porphyry copper deposits, copper values tend to be higher in the potassic than in the phyllic zone.

Sulfide mineralogy is usually simple. Enargite and other sulfosalts are very rare and bornite is as common in some deposits as chalcopyrite. Molybdenite is the only molybdenum mineral observed. Tungstates occur rarely, as at Gaspe. Stannite has been uniquely noted at Catheart.

Some stockwork deposits are on or near and may be genetically related to major strike-slip faults, but no unique fault trend has been identified. At the Gaspe deposit, the major controlling fault strikes east-west, whereas at Woodstock it trends nearly north-south. The variability of trends for major faults does not seem to influence internal structure found in associated deposits.

In most Appalachian orogen porphyry copper deposits, regardless of the trend of major regional structures, mineralization is mostly limited to fractures or joints trending within N30W to N10E. Fractures with other trends are less strongly developed and weakly mineralized as, for example, the east-west set at Catheart. Mineralized fractures are invariably tensional. The persistence of trends that range in azmuth from N30W to N10E in intrusions from earliest Cambrian to Devonian suggests restricted limits for intermittent, repetitive regional tectonic stresses for that portion of the Appalachian orogen studied. Also, the dominance of the veinlet trend in each intrusion, regardless of age, suggests that fractures are related to regional tectonics, except for those that formed uniquely in the habitat of the individual intrusion (e.g., cooling joints and collapse features).

Further evidence supporting the tectonic orogen or tensional fractures can be seen at specific deposits. At Mariner, N10E trending fractures carry ore minerals within the intrusion. In adjacent volcanics, similar trending fractures are filled with quartz or calcite only. At Evandale, the Hampstead shear veinlet trend extends from the stock into intruded Ordovician slates and related fractures cut the circular fractures near the margin of the stock. It seems logical that uniformly trending mineralized fractures in both stock and intruded rock owe their origin to some event unrelated to the intrusion and cooling of pluton.

Some metal zoning, with separation of molybdenum and copper, is apparent in most deposits. Zoning at Gaspe is vertical and the molybdenum:copper ratio increases with depth, whereas zoning at Rencontre East and elsewhere is developed laterally as well. Few porphyries in the Appalachians have peripheral zones of lead and zinc deposits around the mineralized intrusive in contrast to younger deposits elsewhere.

Supergene Concentrations: Minor supergene copper concentrations occur where glaciation has not been severely erosive and where geochemical conditions permitted migration of ionic copper in ground water. Prolonged but generally superficial Quaternary glaciation of the northern Appalachians may have destroyed most supergene copper sulfide concentrations in porphyry copper deposits. Existence of minor supergene mineralization at Woodstock and more rarely at other deposits in weakly glaciated areas indicates that glaciation may not have been a uniquely controlling factor. The geochemistry of host rocks was also effective in regulating movement of ionic copper in ground water. Those porphyry copper deposits exhibiting strong development of a potassic alteration zone coupled with low pyrite content were essentially too rich in such reactive silicates as orthoclase and biotite to have permitted excess hydrogen ion concentrations to form. Because copper as sulfate moves like ions in acid solutions, the lack of oxidizable sulfide minerals in most potassic zones fixed the copper in place during weathering. These results may be seen especially well at Mariner and Clarks Lake. Oxidation largely fixed copper in oxide minerals at Gaspe because metasediments reacted with sulfate waters to neutralize acid. Thus, instead of a prominent supergene blanket large areas of oxide copper minerals were formed.

At Woodstock the presence of as much as 5% total sulfides and erratic sericitization of feldspars were more conducive to conditions favoring ionic migration. Even here, however, supergene copper concentrations are preglacial and conceivably could be pre-Mesozoic, as pre-Carboniferous intrusions in this area show deep weathering below the Carboniferous formations. Coincidentally, evidence now suggests that glaciation near Woodstock was minimally erosional. Even with an optimum environment, however, supergene copper enrichment appears to be insignificant by Cordilleran standards. Pyrite at Catheart may have been sufficient to have promoted supergene processes, but severe glaciation in that area may have removed the concentrations significantly.

Age of Deposits: Radiometric age data for Appalachian orogen porphyry copper deposits are neither abundant nor free of controversy. In this section some of the more significant dates will be summarized.

Age determinations in and near the Catheart deposit in Maine are controversial. The Catheart stock appears to be a late phase of or a separate intrusion into the Attean quartz monzonite. Stratigraphic evidence indicates that the Attean is Ordovician (Albee and Boudette, 1972). Boone, et al. (1970) quote a K-Ar age of 433 and an Rb-Sr age of 457 m.y. on mica from the selvage zone of a molybdenum vein at Catheart. On the other hand, Albee and Boudette (1972) supply a 360 m.y. date for the nearby intrusion at Jackman, hence the Catheart intrusion possibly could be this age also. Schmidt (1974) presents evidence suggesting that Catheart may not be a deeply eroded deposit, while Livingston (1975) has obtained an Rb-Sr date of 320 m.y. on potassic alteration exposed in drill core. The conflict in dates and depth of erosion is yet to be resolved. An Ordovician age for Catheart is preferred on the strength of regional geologic relationships, however.

The Sally Mountain mineralization is also believed associated with a plug that is a late differentiate of the Attean quartz monzonite. If this interpretation is correct, mineralization there would be Ordovician based on regional stratigraphic evidence.

The Evandale stock is dated at 364 m.y. by Ruitenberg (1972) but a copper-molybdenum-bearing shear zone cuts it. Mineralization is undated but could be 364 m.y. or younger.

A K-Ar date of 342 m.y. has been determined on an intrusion near Rencontre East, so the porphyry intrusion could possibly have this date also. Mariner appears to be dated at 580 m.y., or the beginning of Cambrian.

Gaspe, the only producing mine, has been given a 350 m.y. K-Ar date as well as a 346 m.y. K-Ar date (Lamarche, 1973). This is the youngest porphyry copper-type mineralization thus far dated in the Appalachian orogen and suggests that the youngest deposits may be the most economically promising.

Although controversial and inconclusive at present, the evidence suggests that porphyry copper mineralization commenced in

Cambrian, appeared again in Ordovician, and again well into Devonian.

Evidence for Deep Erosion

As a group, porphyry copper deposits of the Appalachian orogen have characteristics and associations that distinguish them from similar deposits in the Cordilleran and Andean orogens. These features are assumed to indicate a greater depth of erosion for Appalachian deposits.

Most significant in the setting of Appalachian porphyries is the absence in all but a few deposits of an extrusive volcanic sequence genetically associated with mineralized pluton. A volcanic-plutonic association exists for many Cordilleran and Andean porphyry copper deposits. Volcanics that should have existed in the Appalachians are assumed to be missing because erosion removed them.

Texture of the intrusion is not porphyritic in half the Appalachian examples, although in Cordilleran deposits the vast majority are porphyritic. Lack of porphyritic texture is believed to reflect greater depth of erosion, which left exposed a deep-seated part of the intrusion.

Lead and zinc sulfides are commonly found in a peripheral zone around Cordilleran and Andean porphyry copper deposits but only rarely in association with Appalachian deposits. Their absence from most Appalachian examples is believed due to a level of exposure where such deposits *bottomed out* and were eroded away. In many known Cordilleran porphyry copper deposits, adjacent lead-zinc mineralization does not persist to the same depth as copper, and the simplest explanation for their absence from the Appalachians is greater depth of erosion.

Weakness or absence of a phyllic zone in Appalachian deposits is taken as evidence that erosion has removed all or much of the zone and that we are actually seeing the roots of a porphyry system. The weakness of pervasive pyritization is similarly interpreted. Presence of more ordered silicates (microcline instead of orthoclase) is compatible with greater depth of formation.

Thompson, et al. (1968) infer a depth of erosion of 15.5 km near Catheart from a metamorphic mineral triple point that is not isochronous with the Catheart intrusion but may suggest the depth of burial during lower Paleozoic.

CONCLUSIONS

Porphyry copper deposits of the Appalachians formed during periods of apparent distension (see Fig. 23, modified after Williams, et al., 1972), although most deposits developed immediately following the Middle Devonian compressive event when tectonic conditions most favored development of extensional fractures at the site of the porphyry copper deposits. Periods of compressive tectonics proved unfavorable for widespread development of porphyry copper deposits.

Well developed potassic zones are more completely exposed in Appalachian orogen deposits than elsewhere. Ferromagnesian silicates accompanying potassic feldspar may be either epidote, epidote-chlorite, chlorite, biotite-chlorite, or biotite. Possibly the epidote formed in response to postmineral metamorphism, but it is more conspicuous in the Appalachian orogen than in deposits elsewhere. Chlorite and biotite are hydrothermal.

Appalachian porphyry copper deposits developed in what is essentially an ensialic orogen. Well developed potassic alteration zones are most likely to be found in this environment and low gold:copper and high molybdenum:copper ratios of most ores (in comparison to Cordilleran orogen deposits) may also reflect this crustal setting.

REFERENCES AND BIBLIOGRAPHY

Albee, A. L., and Boudette, E. L., 1972, "Geology of the Attean Quadrangle, Maine," Bulletin No. 1297, US Geological Survey.

Anderson, F. D., 1968, "Woodstock, Millville and Coldstream Map Area," Memoir No. 353, Geological Survey of Canada.

Beg, M. A., 1975, "The Catawaba Pluton," *Abstracts*, Geological Society of America, p. 469.

Bell, A. M., and Scott, F. J., 1954, "Alteration Associated with Gaspe," *Economic Geology*, Vol. 49, pp. 501-515.

Boone, G. M., 1962, "Origin of Syenite, Deboullie District, Maine," *Bulletin*, Geological Society of America, Vol. 73, pp. 1451-1476.

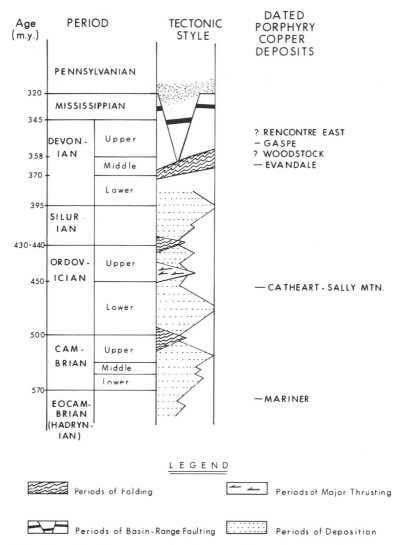

Fig. 23. Correlation chart. Porphyry copper deposits appear from the Hadrynian to Upper Devonian in periods of distensional tectonics. Evandale and Woodstock are posttectonic, although Middle Devonian (*modified after Williams, et al., 1972*).

Boone, G. M., Boudette, E. L., and Moench, R. H., 1970, "The Rangely Lakes Area," *Guidebook*, G. M. Boone, ed., New England Intercollegiate Geology Conference, pp. 1-24.

Burchfiel, B. C., and Davis, G. A., 1972, "Structural Framework and Evolution of the Southern Part of the Cordilleran Orogen, Western United States," *American Journal of Science*, Vol. 272, pp. 97-118.

Cady, W. M., 1972, "Are the Ordovician Northern Appalachians and the Mesozoic Cordilleran Systems Homologous?" *Journal of Geophysical Research*, Vol. 77, p. 8306.

Clark, K. F., 1972, "Stockwork Molybdenum Deposits in the Western Cordillera," *Economic Geology*, Vol. 67, p. 731.

Cook, R. B., 1971, "The Conner Stock, North Carolina," *Economic Geology*, Vol. 66, p. 1003.

Cormier, R. F., 1972, "Radiometric Ages of Granitic Rocks, Cape Breton Island, N.S.," *Canadian Journal of Earth Sciences*, Vol. 9, p. 1074.

Cotelo Neiva, J. M., 1972, "Tin-Tungsten Deposits and Granites from Northern Portugal," *24th International Geological Congress*, Sec. 4, pp. 282-291.

Dewey, J. F., 1969, "Evolution of the Appalachian/ Caledonian Orogen," *Nature,* Vol. 222, pp. 124-129.

Fortier, Y. O., and Hodgson, J. H., 1972, Second Report, Canadian Geodynamics Subcommittee, p. 35.

Helmstaedt, H., and Tella, S., 1973, "Evidence for Avalonian Deformation," *Abstracts,* Geological Society of America, Vol. 5, No. 2.

Hollister, V. F., Barker, A. L., and Potter, R. R., 1974, "Porphyry Deposits of the Appalachians," *Economic Geology,* Vol. 69, No. 5.

Howie, R. D., and Cumming, L. M., 1963, "Basement Features of the Canadian Appalachians," Bulletin No. 89, Geological Survey of Canada.

Irwin, W. P., and Coleman, R. G., 1972, "Alpine Type Ultramafic Rocks," Miscellaneous Field Investigation No. MF 340, US Geological Survey.

Kerr, P. F., 1946, "Tungsten Mineralization in the U.S.," Memoir No. 15, Geological Society of America, p. 242.

Killeen, P. L., and Newman, W. L., 1965, "Tin in the United States," Mineral Investigation No. MR 44, US Geological Survey.

King, R. U., 1970, "Molybdenum in the United States," Mineral Investigation No. MR 55, US Geological Survey.

Kirkham, R. V., and Soregaroli, A. E., 1975, "Preliminary Assessment of Porphyry Deposits in the Canadian Appalachians," Paper 75-1, Geological Survey of Canada, Pt. A, p. 249.

Lamarche, R. Y., 1973, Private Communication.

Livingston, D., 1975, Private Communication.

Lowell, J. D., and Guilbert, J. M., 1970, "Lateral and Vertical Alteration-Mineralization Zoning in Porphyry Copper Deposits," *Economic Geology,* Vol. 65, pp. 373-408.

McAllister, A. L., and Lamarche, R. Y., 1972, "Mineral Deposits of Southern Quebec and New Brunswick," *Guidebook A58,* 24th International Geological Congress.

Morris, W. A., 1976, "Transcurrent Motion in the Appalachians," *Canadian Journal of Earth Sciences,* Vol. 13, pp. 1236-1244.

Naylor, R. S., 1968, "Origin and Regional Relationships of the Core-Rocks of the Oliverian Domes," *Studies of Appalachian Geology,* E-An Zen, et al., eds., John Wiley & Sons, New York, NY.

Pajari, G. E., et al., "The Age of Acadian Deformation in New Brunswick," *Canadian Journal of Earth Sciences,* Vol. 11, pp. 1309-1313.

Potter, R. R., 1970, "Metallogeny and Characteristics of Sulfide Deposits in the Appalachian Region," Annual Meeting Abstracts, Canadian Institute of Mining and Metallurgy.

Potter, R. R., 1973, "New Brunswick Mineral Potential," 41st Annual Meeting, Prospectors and Developers, Toronto, Ont., Canada, April 1973.

Rogers, J., 1970, *Tectonics of the Appalachians,* John Wiley & Sons, New York, NY, p. 271.

Ruitenberg, A. A., 1969, "Mineral Deposits in Granitic Intrusions," Investigation No. 9, New Brunswick Dept. of Natural Resources.

Schmidt, R. G., "Alteration in the Catheart Mo-Cu Deposit, Maine," *Journal of Research,* US Geological Survey, Vol. 2, pp. 189-194.

Sierra, J., Ortiz, A., and Burkhalter, J., 1972, "Metallogenic Map of Spain," *24th International Geological Congress,* Sec. 4, pp. 110-121.

Sillitoe, R. H., and Sawkins, F. J., 1971, "Origin of Copper-Bearing Tourmaline Breccia Pipes, Chile," *Economic Geology,* Vol. 66.

Smith, B. L., 1953, "Fluorite Deposits of Long Harbour, Fortune Bay," Report No. 2, Geological Survey of Newfoundland.

Snelgrove, A. K., and Baird, D. M., 1953, "Mines and Mineral Resources of Newfoundland," Information Circular No. 4, Dept. of Mines and Resources, Geological Survey of Newfoundland, St. Johns, Nfld., Canada.

Thompson, J. B., et al., 1968, "Nappes and Gneiss Domes in West-Central New England," *Studies of Appalachian Geology,* E-An Zen, et al., eds., John Wiley & Sons, New York, NY.

Tupper, W. M., and Hart, S. R., 1961, "Minimum Age of the Silurian in New Brunswick, Canada," *Economic Geology,* Vol. 52, pp. 150-168.

Vokes, F. M., 1963, "Molybdenum Deposits of Canada," Economic Geology Report No. 20, Geological Survey of Canada.

Wanless, R. K., 1969, "Isotopic Age Map of Canada, 1969," Map No. 1256A, Geological Survey of Canada.

Wanless, R. K., et al., 1968, "Age Determinations and Geological Studies," Paper No. 67-2, Geological Survey of Canada.

White, D. E., 1940, "The Molybdenum Deposits at Rencontre East, Newfoundland," *Economic Geology,* Vol. 35, pp. 967-995.

Williams, H., Kennedy, M. J., and Neal, E. R. W., 1972, "The Appalachian Structural Province," *Variations in Tectonic Styles in Canada,* Special Paper No. 11, Geological Association of Canada.

Worthington, J., 1973, Private Communication.

Wynne-Edwards, H. R., and Hasan, Z-ul, 1970, "Intersecting Orogenic Belts Across the Atlantic," *American Journal of Science,* Vol. 268, pp. 298-308.

Zen, E-An, et al., eds., 1968, *Studies of Appalachian Geology,* John Wiley & Sons, New York, NY.

Zietz, I., and Zen, E-An, 1973, "Northern Appalachians," *Geotimes,* Vol. 18, No. 2.

3

Porphyry Copper Deposits of Alaska

CONTENTS

INTRODUCTION

This chapter summarizes porphyry copper deposits within the State of Alaska. Prospecting for porphyry copper-type deposits in Alaska germinated during the 1940's, grew slowly during the '50's, boomed during the '60's, and matured in the '70's. Although large expenditures have been made in the search for porphyry-type deposits, most accrued data lies dormant in company files. However, an adequate base exists in the literature for a general synthesis of areas that contain porphyry prospects. To simplify the summary, porphyry molybdenum deposits as defined by Clark (1972) are omitted.

Description of deposits in their geologic setting permits some correlations between deposit characteristics and crustal environment. These correlations allow speculation about the genesis of the deposits. The summary also attempts to group deposits on the basis of general mineralogic characteristics because diorite model deposits are separated from quartz phenocryst-bearing hosts.

Data presented in this chapter summarize observations by the author, published descriptions of specific areas and deposits as noted in the bibliography, and comments by various government geologists. E. M. Mac-Kevett, Jr., D. H. Richter, and H. C. Berg of the US Geological Survey have been particularly helpful. Because few comprehensive descriptions of individual porphyry copper deposits exist for this area, much of the data presented is new to the literature. Details of structure, isotopic geochemistry, and petrogenesis for individual deposits are largely lacking and even the general geology of some belts is not well understood.

Details on petrography of Alaskan porphyry copper occurrences are inadequate for construction of ternary diagrams (e.g., Figs. 36a, b, pp. 103, 104, for the northern Cordilleran orogen). Petrographic descriptions summarized in the text are largely from published descriptions whose sources are cited.

Porphyry copper-type deposits in Alaska include deposits with pervasive pyrite that contain disseminated copper (but not necessarily coproduct molybdenum). Available data preclude differentiating the deposits by size. No Alaskan porphyry copper deposit is now in production but this is believed to be, in part, a function of economics. In some cases available geologic information is inadequate for classifying deposits. Some deposits included with porphyry copper may, with further exploration, be demonstrated to be some other genetic type.

Alaska has three distinct porphyry copper provinces: the Hogatza plutonic belt, the Continental Margin belt, and the Interior belt. Each belt contains deposits with characteristic geologic features distinct from those of other belts. Fig. 24 shows the continental margin belt to extend inland from

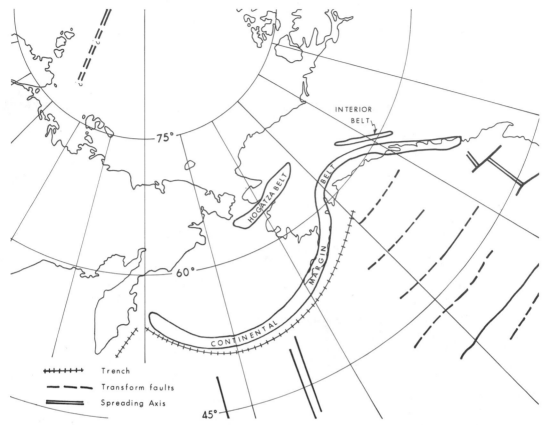

Fig. 24. Porphyry copper belts of Alaska. Porphyry copper occurrences are more numerous in three areas of Alaska: along the continental margin, in the interior, and coincident with the Hogatza plutonic belt. Deposits in each belt have distinct characteristics that set the belt apart. Other isolated deposits are known outside these belts (*base map modified from US Geological Survey Map No. MF-340*).

the southern shore of Alaska, while the interior belt lies just to its north in the Yukon. The Hogatza belt is east and south of the Seward Peninsula.

Lathram and Gryc (1972) include the continental margin and interior belts in their study of metallogenic provinces inferred from space studies. These belts are also implied in the map of metal provinces of Alaska (Clark, et al., 1974).

Porphyry copper deposits in the Hogatza plutonic belt have ages of about 80-82 m.y. (Miller, 1972). They are part of a tourmaline-bearing alteration province and are associated with intrusions dominantly quartz monzonitic in composition. Their alteration zones follow the Lowell and Guilbert (1970) model fairly closely. Table 4 summarizes

characteristics of deposits of the Hogatza belt. Fig. 25 shows this belt to lie in a Mesozoic back-arc province.

Porphyry copper deposits in the continental margin belt are Tertiary, Mesozoic, or Paleozoic in age; are associated with dioritic or quartz-dioritic to quartz-monzonitic rocks; and contain alteration assemblages generally free of tourmaline. Their alteration suites may be potassic or sodic and may diverge widely from the Lowell and Guilbert (1970) model. Tables 6-10 summarize characteristics of deposits in the continental margin belt. Fig. 25 depicts this belt as coextensive with Tertiary and Mesozoic arc-trench sequences.

Porphyry-type deposits in the interior belt are most commonly Latest Cretaceous or Early Tertiary in age, quartz monzonitic in

Table 4. Principal Porphyry Copper Prospects of the Hogatza Plutonic Belt

Name	Quadrangle	Location	Pluton			Alteration Zoning Sequence From Center	Pyrite Zone		Structure	Reference
			Type	Age, m.y.	Host Rock		Size, m x 10³ (x 10³ ft)	% in Phyllic Zone		
West Cape	St. Lawrence	63°28'; 171°32'	Qtz Mon Por	Cret	Cret Vol	Arg Prop	1.2 x 1.2 (4 x 4)	3	Stockwork	Patton & Csejtey, 1971
Granite Mt.	Candle	65°20'; 161°25'	Qtz Mon Por	Cret	Cret Vol	Phy-Arg-Prop	1.8 x 0.9 (6 x 3)	5	Stockwork	Miller & Elliott, 1969
Indian Mt.	Hughes	66°05'; 153°47'	Qtz Mon Por	81.5	Cret Vol	Phy-Arg-Prop	1.5 x 1.2 (5 x 4)	4/four	Breccia	Miller & Ferrians, 1968
Zane Hill	Hughes	66°20'; 156°05'	Qtz Mon Por	81.9	Cret Vol	Phy-Arg-Prop	1.5 x 0.9 (5 x 3)	3	Stockwork	Miller & Ferrians, 1968
Purcell Mt.	Hughes	66°15'; 157°30'	Qtz Mon Por	80	Cret Vol	Arg-Prop	1.8 x 0.9 (6 x 3)	4	Stockwork	Miller & Ferrians, 1968
Chandalar	Chandalar	67°32'; 148°15'	Grdr	370?	Dev Meta	Skarn	—	—	Skarn	Brosge & Reiser, 1964

Cret: Cretaceous   Grdr: Granodiorite   Prop: Propylitic
Qtz: Quartz   Vol: Volcanics   Arg: Argillic
Mon: Monzonite   Dev: Devonian   Phy: Phyllic
Por: Porphyry   Meta: Metamorphics   Pot: Potassic

Table 5. Principal Porphyry Copper Prospects of the Interior Belt

Name	Quadrangle	Location	Pluton			Alteration Zoning Sequence From Center	Pyrite Zone		Structure	Reference
			Type	Age, m.y.	Host Rock		Size, m x 10³ (x 10³ ft)	% in Phyllic Zone		
Casino	Snag	62°44'; 138°50'	Qtz Mon Por	70	Yukon Terrane	Pot-Phy-Arg-Prop	2.4 x 1.8 (8 x 6)	6	Breccia	CIM Spec. Vol. 11, 1971
Co	Snag	62°40'; 138°30'	Qtz Mon Por	Cret	Yukon Terrane	Pot-Arg-Prop	0.9 x 0.9 (3 x 3)	3	Stockwork	Templeman-Kluit, 1973
Cockfield	Snag	62°39'; 138°28'	Qtz Mon	Cret	Yukon Terrane	Arg-Prop	0.9 x 0.9 (3 x 3)	3	Stockwork	Templeman-Kluit, 1973
Dennis	Tanacross	63°23'; 142°27'	Qtz Mon Por	Cret	Yukon Terrane	Pot-Phy-Arg-Prop	2.4 x 1.2 (8 x 4)	5	Stockwork	Hollister, et al., 1974
Taurus	Tanacross	63°29'; 141°20'	Qtz Mon Por	Cret	Yukon Terrane	Pot-Phy-Arg-Prop	2.4 x 1.2 (8 x 4)	5	Stockwork	Hollister, et al., 1974
Tok	Tanacross	63°20'; 142°30'	Rhy Por	Cret	Yukon Terrane	Phy-Arg-Prop	1.2 x 1.2 (4 x 4)	4	Stockwork	Hollister, et al., 1974
Mt. Nansen	Carmacks	62°03'; 137°09'	Rhy Por	Cret	Yukon Terrane	Phy-Arg-Prop	1.5 x 1.2 (5 x 4)	4	Stockwork	Hollister, et al., 1974

Cret: Cretaceous   Grdr: Granodiorite   Prop: Propylitic
Qtz: Quartz   Vol: Volcanics   Arg: Argillic
Mon: Monzonite   Dev: Devonian   Phy: Phyllic
Por: Porphyry   Meta: Metamorphics   Pot: Potassic

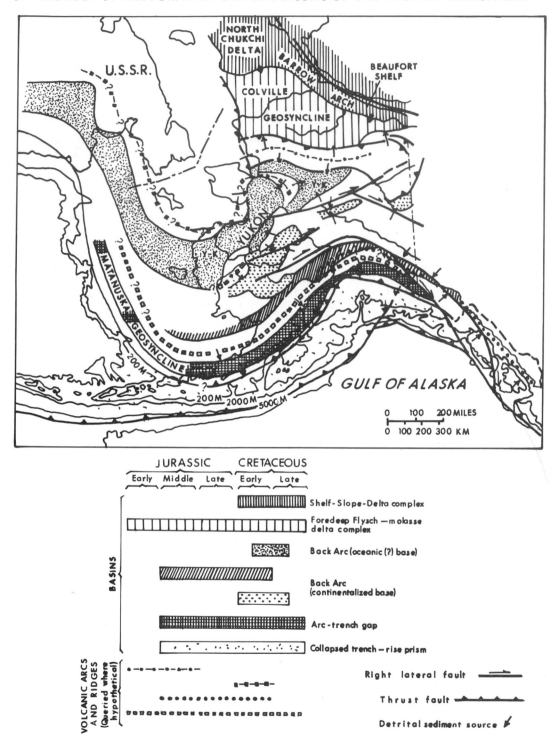

Fig. 25. Mesozoic plate tectonics of Alaska. Tectonic framework of Alaska during the middle and late Mesozoic. As can be seen from Fig. 24, porphyry copper deposits form belts parallel to the Mesozoic tectonic fabric (*modified after Grantz and Kirschner, 1975*).

composition, and bounded by more substantial pyrite halos. Alteration zones in these deposits tend to follow the Lowell and Guilbert (1970) model very closely. Interior belt deposits are summarized in Table 5. This belt, as shown in Fig. 25, is in a cratonic setting.

Isolated deposits that occur outside the three provinces are mainly Tertiary in age.

GENERAL GEOLOGY

As far as is now known, porphyry copper deposits formed as an integral part of the tectonism and magmatism of the Pennsylvanian and younger periods. This summary of geology therefore will emphasize the post-Mississippian geology of Alaska.

Excellent descriptions of the geology of porphyry copper belts appear in *Arctic Geology* (1973), Berg. et al. (1972), or Hollister, et al. (1975). This introduction generally mentions broader tectonic features and their relationships to Alaskan porphyry deposits; further detailed summaries of each belt follow.

Plate movements recorded in the Pacific (Naugler and Wageman, 1973) indicate active subduction under Alaska's southern coast on an intermittent basis at least since the Early Jurassic. Onshore geologic studies (e.g., Moore, 1973) suggest that southern Alaska includes classic examples of continental accretion of eugeosynclinal material caused by underthrusting of oceanic crust beneath an island arc. Upper Paleozoic to Jurassic igneous activity included prominent basic volcanism along the southern Alaskan coast. Magma generation for Jurassic batholithic rocks (175 to 155 m.y. interval) appears to have occurred along or above a Mesozoic subduction zone. Porphyry copper deposits have not yet been found in Alaska with dates in this interval.

Magma generation for smaller Upper Cretaceous-Tertiary plutons is given a more complex origin by Reed and Lanphere (1973), with no obvious simple connection between subduction and magmatism. Porphyry copper deposits do occur with some of these younger and smaller plutons. Rarity of porphyry copper mineralization coeval with large-scale batholithic intrusion implies that tectonic factors favoring Sierra Nevada-type batholiths do not necessarily favor porphyry copper magmatism or mineralization. Rather than rapid subduction that appears to coincide with large-scale batholith intrusion (Larson and Chase, 1972), porphyry copper magmatism apparently favors slow subduction or strike-slip tectonics. On the other hand, Armstrong, et al. (1976) provide dates on porphyry copper deposits in the Aleutians that are concordant with the active, extensive, Late Upper Tertiary arc-trench system. A good case can be made for simultaneous subduction and porphyry mineralization for these deposits.

The largest number of prospects and largest developed reserves are in the continental margin belt. Mesozoic crustal setting for each belt is shown in Fig. 25.

HOGATZA PLUTONIC BELT

Introduction: The Hogatza plutonic belt was described by Miller, Patton, and Lanphere (1966) to include the alkaline intrusive province on and to the east of Seward Peninsula. Here it is enlarged to include copper-bearing mineralized intermediate intrusions that extend intermittently from St. Lawrence Island generally eastward into the Yukon-Koyukuk province (Fig. 26). The US Geological Survey 1:250,000-scale quadrangles involved are St. Lawrence, Candle, Norton Bay, Kateel River, Hughes, Shungnak, and Selawik.

A porphyry copper prospect occurs on St. Lawrence Island in the west but none have been found on Seward Peninsula. Geologically, porphyry copper-type deposits in the eastern portion of this belt are limited on the south by the Kaltag fault and appear bounded on the north by the Kobuk fault. No substantial porphyry copper deposits are known east of the intersection of these two faults, although tin, tungsten, and molybdenum mineralization associated with epizonal intrusive activity has been reported to the east. Similarly, tin, tungsten, beryllium, and molybdenum mineralization occurs spatially associated with intrusions on Seward Peninsula of the same general age as the porphyry copper deposits of the Hogatza plutonic belt.

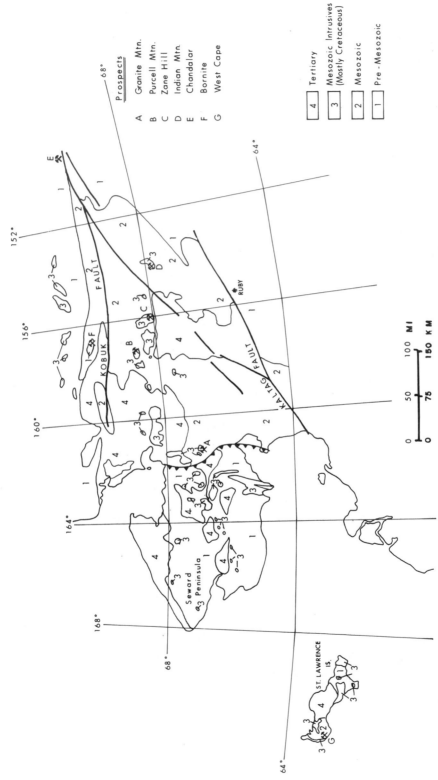

Fig. 26. Outline geology of the Hogatza plutonic belt, Alaska. Porphyry-type mineralization present within the Hogatza plutonic belt extends northeastward from St. Lawrence Island. On the Seward Peninsula it consists of porphyry lithophile element occurrences. East of Seward Peninsula, porphyry copper deposits again reappear between the Kaltag and Kobuk faults (*modified from numerous US Geological Survey sources*).

Where these lithophile-element deposits occur, porphyry copper prospects are absent. Sainsbury (1969) describes fluorite and fluorsilicates as alteration products associated with Seward Peninsula molybdenum-lithophile porphyry deposits, a feature that further distinguishes them from porphyry copper deposits occurring to the east and west.

Porphyry copper deposits south of Kobuk fault lie south of extensive base-metal deposits in the greenschist facies volcanic and Devonian carbonate rocks of the Amblar district. Some of these strataform deposits, e.g., Bornite (Figs. 26 and 27), may contain very large reserves of zinc and copper-bearing material. Their Paleozoic age (Bornite is Middle Devonian, whereas other deposits—e.g., Arctic Camp and Picnic—occur within metavolcanic sequences appearing to be Paleozoic) clearly separates them from any genetic tie to younger porphyry copper deposits, however.

The Chandalar copper-bearing skarns (Figs. 26 and 27) also have some copper ore potential and a pluton near them has a zircon age of 270 m.y. (Brosge and Reiser, 1964). The source of copper in the skarns has not been clearly identified and the deposits may be polygenetic. In any case they are not clearly porphyry copper-type deposits and therefore do not have a close bearing on the genesis of Cretaceous porphyry copper deposits of the Hogatza plutonic belt.

In summary, the Hogatza porphyry copper belt extends west from the intersection of Kaltag and Kobuk faults to and including St. Lawrence Island but excluding that part of Seward Peninsula underlain by lower Paleozoic and Precambrian rocks.

Geologic Setting: Fig. 26, after Miller (1970), Patton (1970, 1973), the US Geological Survey Tectonic Map, and Sainsbury (1969), shows general geology of the Hogatza plutonic belt. However, this summary emphasizes that portion of the Hogatza plutonic belt underlying and hosting porphyry copper deposits. Geology of the Hogatza plutonic belt shown in Fig. 26 is largely simplified into a pre-Mesozoic and Mesozoic

host for both alkalic and calc-alkalic epizonal and mesozonal intrusions. Undivided Tertiary sedimentary and volcanic rocks are shown on this figure generally on and near the peninsula. To emphasize porphyry copper occurrences, only the Chandalar and Bornite nonporphyry deposits are shown in Fig. 26.

Because most deposits occur in the wedge between the Kobuk and Kaltag faults, more detailed descriptions are given for this area. Except for the St. Lawrence Island occurrence, porphyry copper deposits occur in a wedge-shaped depression of Cretaceous and Tertiary volcanic, volcanoclastic, and sedimentary rocks bordered on the west, north, and southeast by a metamorphic complex containing Permian or older formations. Patton (1973) establishes the wedge as a highly mobile terrane subjected to repeated volcanism and plutonism during the Cretaceous and Tertiary. Greater geologic detail within the Kobuk-Kaltag fault wedge is presented in Fig. 27. The oldest rocks exposed within the wedge are a thick section of earliest Cretaceous marine andesitic rocks that either overlie the Rampart group (Brosge, et al., 1969) or older continental crust. In either case, the thick section of more basic marine volcanic rocks in the wedge forms a foundation through which mineralized porphyry copper intrusions penetrated. Development of the wedge through lateral tectonic displacement of sialic crustal plates (Patton, 1973) may have favored development of oceanic crust as the continental crust moved. A viable alternative to the displaced-plates hypothesis suggests that the continental crust in the wedge subsided and *thinned* in a rifting environment.

Marine clastic and volcanoclastic sedimentary rocks deposited in two Albian and Cenomanian troughs [as deep as 7620 m (25,00 ft)] overlie the earliest Cretaceous marine volcanic rocks. Older Cretaceous subsilicic stocks with K-Ar ages of about 100-105 m.y. intruded the marine volcanics. Subarial felsic volcanic rocks of Late Cretaceous and Early Tertiary age are widespread locally and appear to have been

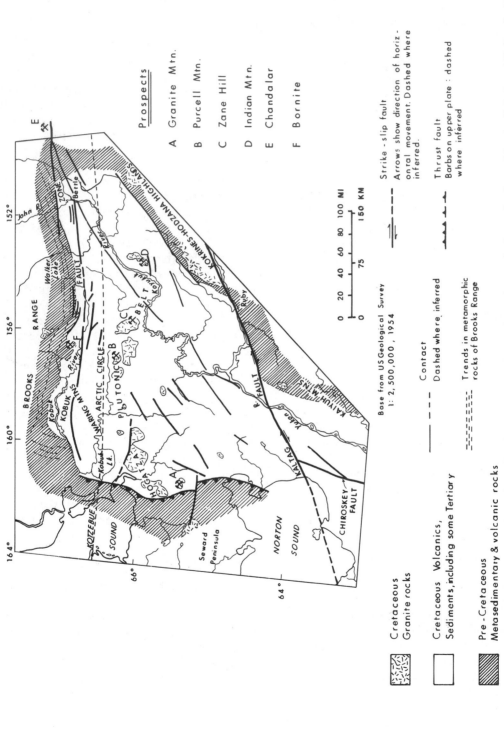

Fig. 27. Geology of the Kobuk-Koyukuk Province. Porphyry copper occurrences are known in the wedge formed by the Kobuk and Kaltag faults and the thrust at the eastern base of Seward Peninsula (*modified after Patton, 1973*).

comagmatic with intrusions in the porphyry copper deposits. Widespread recent loess deposits mask much of the surface.

The Kaltag fault bounding part of the wedge on the south has a strike-slip displacement of 64-129 km (40-80 miles) (Patton, 1973). Old continental crust is brought against Cretaceous marine volcanic rocks in the wedge by this movement. The Kobuk fault bounding the wedge at the north is the locus of major structural discontinuities, but evidence for establishing amount of movement is imprecise. The western margin of the wedge is marked by a poorly exposed thrust zone (Patton, 1973).

Metamorphism of Paleozoic and older rocks preceded, to some extent at least, eruption of Cretaceous marine volcanic rocks. The metamorphism locally reached the amphibolite isograd, apart from local contact thermal effects. K-Ar dating suggests a last resetting of the time clock at about 101 m.y. (Miller, 1970). The thermal event that occurred prior to the Tertiary was the last major regional event. Some movement on major faults postdates the regional metamorphism and all porphyry copper deposits are believed younger than 101 m.y. The deposits therefore are posttectonic.

Table 4 lists characteristics of porphyry copper-type deposits associated with Hogatza plutons. Chandalar skarns are also included in this table.

Crustal conditions on Seward Peninsula where molybdenum-lithophile porphyry deposits occur are different. An apparently thick Precambrian and Paleozoic crust provides a setting not found in adjacent areas where only porphyry copper deposits developed. Although igneous rock for all porphyry deposits remains compositionally within restricted limits and a tourmaline province encompasses both the porphyry copper province and molybdenum-lithophile types of mineralization, a distinction in metallization and alteration is apparent in each deposit type. A somewhat similar relationship between copper-molybdenum porphyries and areas of basic volcanics was noted by Sutherland Brown, et al. (1971) in British Colum-

bia. In that area Sutherland Brown, et al. (1971) report porphyry copper mineralization accompanies calc-alkalic plutons intruding marine volcanic rocks, but porphyry molybdenum mineralization accompanies stocks of the same igneous suite intruding the coastal plutonic complex.

Characteristics of the Porphyry Copper Deposits: Although most intrusions of the Hogatza plutonic belt have K-Ar dates older than 100 m.y. (Miller, 1972), the older, undersaturated plutons having alkalic affinities (Miller, et al., 1966) tend to be unmetallized. Mineralization appears restricted to the youngest period of documented plutonic activity. Plutons associated with mineralization are porphyritic and usually contain several different phases of quartz-bearing igneous rock. However, that phase closest to ore spatially and temporally appears to be quartz monzonitic. Hornblende occurs in each mineralized pluton. Andesine is the average composition of zoned plagioclase and microcline has been identified accompanying orthoclase in some plutons (Miller, et al., 1966). Usually mineralized intrusions are oval (e.g., Indian Mountain), apparently forcefully injected, epizonal, with radiometric dates of 80-82 m.y. (Miller, et al., 1966).

Comparison of porphyry deposits of the Hogatza belt with those of the interior belt shows many petrographic similarities. In each belt quartz monzonite porphyry occurs most commonly with sulfide. However, cratonic conditions (with a thick Precambrian-lower Paleozoic sialic crust) underlying interior belt deposits contrast sharply with a thick surficial section of basic marine volcanic and sedimentary rocks that hosts the Hogatza belt. Island-arc assemblages of the continental margin belt appear to have an upper-crustal composition similar to host rocks of the Hogatza belt, yet petrography of deposits in the continental margin belt is distinct from that of the Hogatza belt. It seems justified to conclude that an old but rifted and distended continental crust underlies volcanics of the Hogatza belt. Petrographic similarities between interior and Hogatza belt deposits may reflect their penetration of similar older continental crust.

Alteration zones shown in Table 4 may be incomplete because of poor outcrop. The summary given for each deposit no doubt could be modified by further exploration. However, zones indicated for each deposit are typical for quartz monzonite intrusions associated with porphyry copper mineralization as compiled in the Lowell and Guilbert (1970) model.

No potassic zone has been identified in the core of any deposit. Destruction of orthoclase during weathering prevents potassic zone recognition in areas of scarce outcrop. Chlorite or biotite could have accompanied orthoclase in such a zone.

Persistence of sericite during weathering, on the other hand, permits easy recognition of a phyllic zone. Strong pyritization and silicification accompany sericite in both Granite Mountain and Indian Mountain. Although pyrite content in the phyllic zone may be weaker in Zane Hill, strong silicification and quartz-sulfide veining occur in this phyllic zone.

An argillic zone is exposed in each deposit shown in Table 4. This may be the product of supergene alteration of pyritized material, but probably some hypogene clay minerals formed an argillic zone peripheral to the phyllic zone. In outcrop the argillic zone is composed of clay minerals and pyrite and is localized between the phyllic and propylitic zones.

A typical chlorite-rich propylitic zone is peripheral to the argillic zone of each deposit. Pyrite may occur erratically with propylitic minerals, but hydrated iron and aluminous silicates typify this assemblage.

Pyrite forms a halo for each deposit, resulting in well defined color anomalies (Miller and Elliott, 1969; Miller and Ferrians, 1968). Size of the pyrite halos shown in Table 4 may be modified by further exploration because outcrop over most deposits is poor and leached capping covers sulfides in each case. Some areas of pyritization have been noted in addition to those listed with individual deposits. These have not been

clearly associated with an outcropping younger intrusion and no significant metallization has yet been found associated with them. They may, however, reflect buried deposits that have not yet been exposed by erosion.

Mineralization is of the copper-molybdenum type. Some have zoned peripheral zinc, lead, silver, and gold veins on their margins. Uranium may be significant in a few deposits (e.g., Zane Hill) as a constituent of disseminated minerals within the pyrite halo. Miller and Ferrians (1968) note that tourmaline occurs with alteration and mineralization in several deposits (e.g., Indian Mountain), and Sainsbury (1969) mentions tourmaline in association with some tin-tungsten-molybdenum deposits on Seward Peninsula. Tourmaline may occur as an alteration product such as rosettes replacing magmatic ferromagnesian silicates in igneous rocks, as vein filling with other silicate and sulfide minerals, as a constituent cementing breccia fragments, or as an apparent magmatic mineral in aplite dikes. Additionally, tourmaline may occur as a fluorine-bearing silicate accompanying topaz and sericite in greisen associated with tin-tungsten-molybdenum mineralization.

Where tourmaline occurs in the phyllic zone of porphyry copper deposits, it is associated with sericite, quartz, and pyrite. It has also been noted in argillic and propylitic zones of some Hogatza belt prospects, and therefore is not a diagnostic alteration mineral.

Tourmaline varies in amount from deposit to deposit, is absent at some deposits, and accompanies others from the Lost River area (Sainsbury, 1969) eastward to Indian Mountain (Miller and Ferrians, 1968). The arc described by the tourmaline province approximately parallels the Aleutian trench 600 miles to the south. Unlike the tourmaline province of the Andean orogen, which is parallel to and about 150 miles away from the present trench, the tourmaline province of the Hogatza belt formed at such a distance from a hypothetical trench off Alaska's south coast that no simple genetic tie may be made between subduction in a Mesozoic trench and tourmalinization. The age of the tourmaline province (80-82 m.y.) also predates the modern Aleutian trench. Since tourmaline has only been reported to occur with Upper Cretaceous quartz-bearing plutons, it appears to be associated with hydrothermal events of a restricted geologic significance in the Hogatza belt.

Exploration of known deposits of the Hogatza belt has not progressed to the stage where any reserves have been developed, nor has the belt itself been exhaustively evaluated. Isolation of the deposits and of the belt itself has inhibited serious past exploration activities. These deposits do not appear affected by glaciation and therefore may have a supergene sulfide zone below exposed leached cappings. Even this added attraction has not provided sufficient incentive to promote adequate exploration of the belt.

INTERIOR BELT

Introduction: A northwest-trending belt of porphyry copper prospects exists in the interior of eastern Alaska and western Yukon (Hollister, et al., 1975). Most prospects in this belt have had pilot exploration programs and therefore the deposits are fairly well known. The more representative deposits are shown in Fig. 28 and summarized in Table 5.

Reserves developed on Casino, the most important deposit thus far discovered, are tentatively placed at 154 million mt (170 million tons) of 0.37% Cu with 0.02% Mo. No other deposit has received such extensive exploration as Casino, however, and all should be considered prospects. Drilling to date generally has shown a very low hypogene copper grade. No deposits in the belt are glaciated and supergene concentrations may be found that will provide economic ore deposits. Leached capping has been found over each prospect. Geologically this belt of porphyry copper deposits is bounded southwest by the Denali fault and northeast by the Tintina fault.

Geology: Excellent summaries of geology within the belt may be found in Templeman-Kluit (1973 and 1976) and Foster (1970).

Berg, et al. (1972) provide some data on post-Devonian geology of the area and Hollister, et al. (1975) outline the belt's geology.

Fig. 28, after various US Geological Survey and Geological Survey of Canada sources, outlines the geology of the Interior belt. Metamorphic terrane north of the Denali fault, referred to as the Yukon Metamorphic Complex (Templeman-Kluit, 1973) or the Yukon Terrane (Berg, et al., 1972) appears to be largely part of the cratonic interior of North America (Berg, et al., 1972). It is mostly composed of metasedimentary rocks. Generally metamorphic grade and probably geologic age of the terrane increases to the north. Close to Denali fault low greenschist facies rocks with lenses of recrystallized limestone locally contain Devonian corals. These rocks appear to grade northward into schists and gneisses of the amphibolite facies that lack diagnostic fossils (Foster, 1970). Neither the metamorphic history nor absolute age of most of the rocks in the Yukon Terrane is well known. A few radiometric dates suggest a number of metamorphic events ranging from Triassic to Lower Cretaceous (190-120 m.y.) and one whole rock date had yielded a 1173 m.y. age (Hollister, et al., 1975). Regional metamorphic events clearly precede development of porphyry copper deposits, as in the Hogatza plutonic belt.

Preliminary radiometric dates on granitic rocks that intrude metasedimentary terrane also suggest a number of periods of plutonic activity. Most batholithic plutons have ages in the 92-86 m.y. range. A Late Cretaceous-Early Tertiary event or events (71-50 m.y., Hollister, et al., 1975) includes stocks that may be associated with porphyry copper deposits. Still older premetamorphic plutonic events are indicated by foliated granitic rocks within the terrane (Templeman-Kluit, 1973). The interior belt of porphyry copper deposits appears genetically associated with the youngest period of documented plutonic activity. No deposits are known to have radiometric dates older than latest Cretaceous, nor does field evidence indicate that older dates should be expected. Pre-Upper

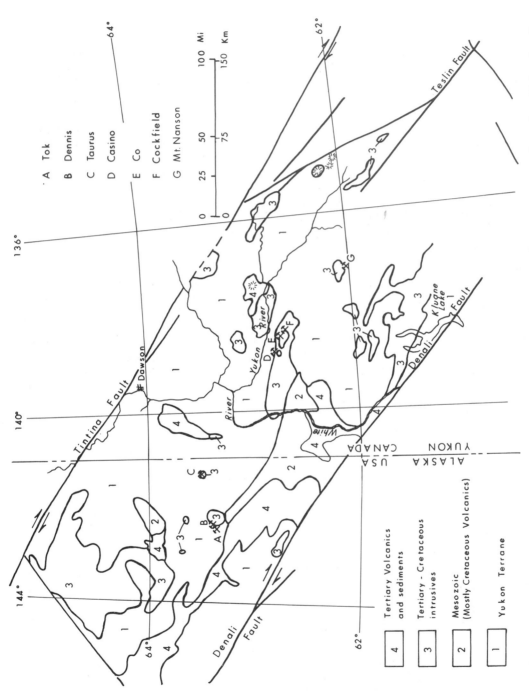

Fig. 28. Interior Belt. The linear trend of porphyry copper occurrences of the Interior Belt of eastern Alaska and the Yukon is strikingly shown in this diagram. On geologic evidence, all are believed to be closely comparable in age to Casino (70 m.y.).

Cretaceous Mesozoic volcanics in the belt (Berg, et al., 1972) apparently do not have porphyry mineralization attendant with epizonal intrusions. Marine volcanism is almost completely absent in Yukon Terrane rocks.

Characteristics of Porphyry Copper Deposits: Porphyry copper deposits of the interior belt have, as can be seen in Table 5, a surprising repetition of features regardless of the size of the alteration halo. Intrusions spatially and temporally closest to mineralization have remarkably constant petrography. Alteration zones from the core outward may be predicted by the Lowell and Guilbert (1970) model. This leads to a repetition of characteristics within interior belt deposits similar to that found in the porphyry copper province in the southern Cordilleran orogen.

Comparison of the geologic environment of interior belt deposits with those of the southern Cordilleran orogen also reveals many similarities. Each has intrusions that penetrated Precambrian or younger metamorphosed sialic crust and radiometric dating of mineralization in one area overlaps that of the other. Therefore, porphyry deposits formed penecontemporaneously in each area with what could be described as a similar cratonic or quasicratonic environment.

Petrography of the mineralized intrusions is fairly simple. Most are epizonal complexes that contain granodiorite and quartz monzonite phases. No microcline has been reported and orthoclase appears to be the only potassic feldspar. Plagioclase is usually zoned with andesine as the average composition (Templeman-Kluit, 1973). Biotite and hornblende are ubiquitous. All plutons closely associated with porphyry mineralization are quartz monzonitic and porphyritic, with euhedral to subhedral quartz phenocrysts characteristic of deposits in this belt. Petrographic similarity between mineralized intrusions of the Hogatza and interior belts has been noted previously. Generally the intrusion has an irregular outline, is epizonal, and appears to be forcefully injected.

Alteration zones noted in Table 5 are fairly well defined due to exposure by trenching and drilling in most deposits. Because surface exposures are poor, the summary given for each deposit in this table may be modified by further exploration, but such changes probably will be inconsequential. Alteration zones indicated for each deposit typify deposits associated with quartz monzonite plutons. Tourmaline is also found in some deposits, e.g., Casino (Godwin, 1975, and *CIM Special Volume* 11, 1971). Where it does occur in a deposit, it is usually found in all alteration zones.

Potassic zones have been identified in most deposits shown in Fig. 28. In these zones secondary orthoclase and biotite are dominant, with only minor pyrite and sericite accompanying major minerals. Chlorite has been reported occasionally and magnetite has also been identified in this zone. Hypogene copper sulfide has been found only occasionally in potassic zones in these deposits.

Phyllic zones are characterized by substantial areas of quartz, sericite, and pyrite surrounding a potassic zone. In some cases (e.g., Casino, Dennis, and Taurus) the phyllic zones are very large (Hollister, et al., 1975). Pyrite content varies from 3 to 6% in this zone. Silicification accompanying phyllic zone minerals is strong enough in some cases to make the ground resistant to erosion so that this zone may form a topographic high (e.g., Tok, Casino). Hypogene copper content, in spite of extensive development of sericitic alteration, is commonly under 0.1% Cu for these deposits.

An argillic zone has been found peripheral to the phyllic zone in each deposit in Table 5. Although supergene argillization resulting from pyrite weathering is extensive in most deposits, a primary argillic zone (as defined in the Lowell and Guilbert model, 1970) has been found in every occurrence. Pyrite accompanies clay minerals, and chlorite commonly appears as a minor constituent. Copper has not been found in the argillic zone of any deposit.

Adjacent to the argillic zone in every deposit is a chlorite-rich, pyrite-poor propylitic zone. The propylitic zone may contain occasional zinc-lead sulfide veins in some metallogenically zoned deposits (e.g., Casino, *CIM*

Special Volume 11, 1971), but rarely contains copper mineralization.

Mineralization in deposits listed in Table 5 appears to be a copper-molybdenum type, with molybdenum zonally concentrated in one part of the deposit. Where present, gold, silver, lead, and zinc are zonally arranged sporadically about the phyllic alteration suite. Copper tends to be erratically distributed through the phyllic zone. Although the belt is primarily a copper-molybdenum province, minor tungsten has been reported in a few deposits (Templeman-Kluit, 1973). Tin- and gold-bearing placers have also been found close to some prospects.

Examples of Interior Belt Deposits: Table 5 indicates that both breccia pipe and stockwork structural types exist in interior belt deposits. One example of each, Taurus and Casino, is summarized as representative of porphyry copper deposits in this belt.

Taurus—Taurus is a deeply weathered stockwork-type deposit that contains—moving outward from a well developed core potassic zone—very large phyllic, argillic, and external propylitic zones. These alteration assemblages are developed in and near a quartz monzonite porphyry intrusion that invades gneisses and schists (Hollister, et al., 1975). Secondary orthoclase, biotite, quartz, and plagioclase characterize the potassic zone. Minor sulfide and prominent magnetite occur as disseminations in this zone, but both copper and molybdenum occur only sparingly.

Pervasive fine-grained sericite intergrown with quartz and pyrite dominates the phyllic zone. Quartz-sericite veinlets bearing copper and molybdenum sulfide occur only in this zone. Kaolin and carbonates are the most important minerals in the argillic zone and both hypogene and supergene copper sulfides are absent. Chlorite is the characteristic silicate of the propylitic zone and calcite occurs abundantly disseminated in the gneiss as a secondary mineral.

Dominant veinlet trend exposed in outcrops is N70W and the pyrite halo extends for 2.5 km (8000 ft) in this direction. The pyrite halo is about 1.3 km (4000 ft) wide.

Pyrite occurs as a dissemination in all alteration zones but is most importantly developed in the phyllic. It also occurs in quartz-sulfide veinlets with or without copper and molybdenum sulfides. Anomalous concentrations of gold have not been found.

Casino—The Casino porphyry copper deposit has been exhaustively described by Godwin (1975). Yukon Terrane schists, gneisses, and quartzites were intruded by the 90 m.y.-old Klotassin granodiorite. Both formations were then intruded by a younger complex (70 m.y.) containing phases ranging in composition from quartz diorite to granite. Dacite and rhyolite intrusions succeeded the younger complex and breccia pipes associated with copper and molybdenum ore minerals and alteration followed. The sequence of intrusion and mineralization was completed with a postmineral cobble breccia equivalent to the pebble breccia at Toquepala, Peru. Including cobble breccia, the area brecciated extends 700 x 400 m (2200 x 1300 ft).

A potassic core zone contains secondary orthoclase, biotite, and tourmaline and lesser amounts of anhydrite, magnetite, ankerite, and hematite within tourmaline breccia pipes. Sulfides are sparse in this zone. Peripheral to the potassic zone are gradational, well developed phyllic and argillic zones dominated by quartz, sericite, tourmaline, sulfide, and montmorillonite clay. Chalcopyrite and molybdenite, with minor sphalerite and ferberite, occur in the phyllic zone. Although the ore minerals appear spatially related to the potassic-phyllic boundary, they occur entirely within the area of tourmalinization. Indeed, only minor hypogene copper sulfide has been found in annular veinlets outside the tourmaline breccia pipe in argillically or propylitically altered rock. The pyrite halo around the pipes is 2.5 x 2.0 km (8000 x 6000 ft).

CONTINENTAL MARGIN BELT

The continental margin belt occupies all southern coastal areas of Alaska, including the Aleutian Islands. This belt is divided into eastern and western sections by Cook Inlet. The western portion of the belt is the site of predominantly Tertiary porphyry copper prospects, whereas the eastern section contains prospects which date from 282 m.y. to Tertiary. As the geology in each area differs fundamentally, each part is discussed separately. Porphyry copper deposits in the western segment are listed in Table 6; those in the east are listed in Tables 8 and 9.

Western Section

Introduction—Geology of the western section is covered in detail by Burk (1965), Reed and Lanphere (1973), and Naugler and Wageman (1973). Porphyry copper in this portion of the continental margin occurs most commonly south of the Denali fault. The area includes part of the Alaska Range, the Aleutian Range, and the Aleutian Islands.

Exploration activity for base-metal deposits has been minimal in this area in the past; hence, few occurrences noted in the western section have been explored. Nor has this section itself been adequately covered in reconnaissance prospecting. Poor weather conditions, restrictions on mineral entry by various government agencies, lack of infrastructure, and isolation have combined to restrict prospecting activity.

Geology—The western part of the continental margin belt is composed largely of arc-trench rocks and their coeval plutons. No rocks in these volcano-plutonic arc complexes are older than upper Paleozoic. Some minor Precambrian and lower Paleozoic sedimentary rocks have been mapped south of and close to the Denali fault. Stocks and small plutons, mostly of Tertiary age, intrude the arc-trench batholith complexes. On the Aleutian arc, Tertiary and Quaternary volcanic rocks cover the older Tertiary. That part of the arc west of Unalaska is not properly a continental margin environment but an island arc in an oceanic setting (Naugler and Wageman, 1973; Marlow, et al., 1973) that has a primarily lower Tertiary development. This portion of the Aleutian arc did not exist in the Mesozoic (Grantz and Kirschner, 1975). Porphyry copper occurrences associated with younger Tertiary intrusions occur in both the peninsular (Unalaska and the northeast) and oceanic crustal or insular

(to the west of Unalaska) environments. The porphyry prospect on Attu lies in the oceanic crustal setting and the peninsular settings occur in the continental margin.

Insular Geology—Rocks exposed west of Unalaska may be divided into three separate unnamed units. The oldest series consists of spilitic and keratophyric lava, breccia, and tuff interbedded with such exhalative products as siliceous and calcareous cherts and cherty argillite. Gabbro intrudes these rocks. The volcanic rocks are now greenstones that have been pervasively albitized. Results of chloritization, epidotization, silicification, and zeolitization are widespread and pyritization is conspicuous in this unit on some islands. Although Marlow, et al. (1973, p. 1556), note that these rocks lack the ultramafics of classic ophiolite assemblages, they are generally accepted to include oceanic crust in a Lower Tertiary volcanic arc association. Volcanic rocks have ages from Eocene to middle Miocene (Burk, 1965).

Plutonic rocks invading the oldest series comprise a middle unit separating basement oceanic crust and arc volcanic rocks from younger Late Tertiary volcanic and sedimentary rocks. K-Ar radiometric dating provides ages of 11.1 and 15.8 m.y. for plutons from Unalaska and Amchitka, respectively. Plutonic rocks range in composition from gabbro to granite, but quartz diorite is most common. Plutonic complexes may have granodiorite and quartz monzonite as younger phases, but volumetrically quartz diorite is most important. Analyses available on these plutonic rocks show the soda-to-potash ratio to be commonly three or four to one.

Generally only the early series is invaded by plutonic rocks and both units are overlain by Late Tertiary vocanic and volcanogenic deposits and sediments. Volcanic rocks of the Quaternary volcanoes cap all older formations. Late Tertiary rocks are distinguished from the early series by absence of marine deposits, deformation, and greenschist facies metamorphism and hydrothermal alteration in older rocks. Andesite apparently comprises about 90% of these Late Tertiary and Quaternary volcanic rocks; the remainder includes basalt, dacite, and rhyolite. Insular porphyry copper deposits appear to be related primarily to the younger age of volcanic activity. They developed in a currently active volcanic arc that parallels the Aleutian trench.

Summarizing briefly, it is clear that periods of voluminous volcanism were confined to Early Tertiary (60-40 m.y.) and Late Cenozoic (8 m.y. to present); but the Aleutians had achieved their present bulk prior to the onset of Late Cenozoic activity. Based on stratigraphic evidence, porphyry copper deposits in this area appear to have ages less than 40 m.y. but, more importantly, less than 15 m.y.

Peninsular Geology—Northeast of Unalaska, the landmass west of Cook Inlet is dominated by the Alaska-Aleutian Range batholith (Reed and Lanphere, 1972, 1973). From near the Denali-Farewell fault system south, the batholith consists of plutonic rocks of mostly Jurassic age comagmatic with arc-trench marine andesitic rocks and volcanoclastic strata. Quartz diorite and granodiorite are the most common intrusive phases. Basement rocks south of 60° may not include significant amounts of pre-Mesozoic crust and rocks south of 58° are largely Tertiary and Mesozoic sedimentary rocks. South of 60°, the basement is probably oceanic crust (Grow, 1973). South of the Alaska-Aleutian batholith, successive arc-trench assemblages provided rocks that now make up the crust (Fig. 25).

Late Cretaceous and Early Tertiary plutonic rocks emplaced between 83 and 58 m.y. cut the older arc-trench systems and those Tertiary rocks deposited on them. The largest intrusions, however, are found closer to the Denali fault. Less abundant younger plutons from 38 to 26 m.y. in age also occur in this area. Some of the latest Cretaceous and Tertiary plutons are associated with extrusive rocks and are considered posttectonic (Reed and Lanphere, 1973). Porphyry copper deposits appear to occur only in Tertiary plutonic groups where they are associated with magma generated beneath both stable platform areas and former geosynclinal regions in which deformation had essentially ceased. Magma generation for porphyry copper deposits associated with some Tertiary plutons is not easily related to a subduction zone (Reed and Lanphere, 1973). On the other hand, Dry Creek (3.3 m.y.) and Pyramid (6.3 m.y.) developed during a period of currently active arc-trench formation. Fig. 29 summarizes the relationship between younger intrusions and the Jurassic batholith in the area closer to the Denali fault.

Arc-Trench Systems—The oldest apparent arc-trench, volcano-plutonic sequence appears to be Early and Middle Jurassic, containing most of the Alaska-Aleutian range plutonic rocks emplaced between 175 and 155 m.y. ago (Reed and Lanphere, 1973). Aeromagnetic data suggest that this Jurassic magmatic arc now visible south of the Denali continues into the Bering Shelf and is called the Matanuska Geosyncline (Fig. 25) by Grantz and Kirschner (1975). The associated trench forms a belt of melange rocks southeast of the old arc. Other younger arc-trench systems have been described but are less perfectly defined. An Upper Jurassic-Lower Cretaceous system is exposed on Kodiak Islland and to the east. Another Upper Cretaceous-Lower Tertiary arc previously described for the Aleutians in the insular geology section is the youngest.

Grow (1973) infers an oceanic crustal basement for the Aleutian arc-trench sequence; it seems likely that porphyry copper-associated magmas penetrating nonbatholithic rocks intruded oceanic crust as well as the trench melange. The host for many porphyry deposits therefore is thought to include basic marine volcanic rocks.

Mineralization in the western part of the continental margin belt is spatially and temporally associated with calc-alkalic intrusions that range from quartz diorite to quartz monzonite. No mineralized diorite or alkalic intrusions asociated with porphyry copper-type mineralization have thus far been recognized in this area, although they are known in the eastern section. In addition to deposits noted in Table 6, a large number of sulfide-caused color anomalies are known that may upon

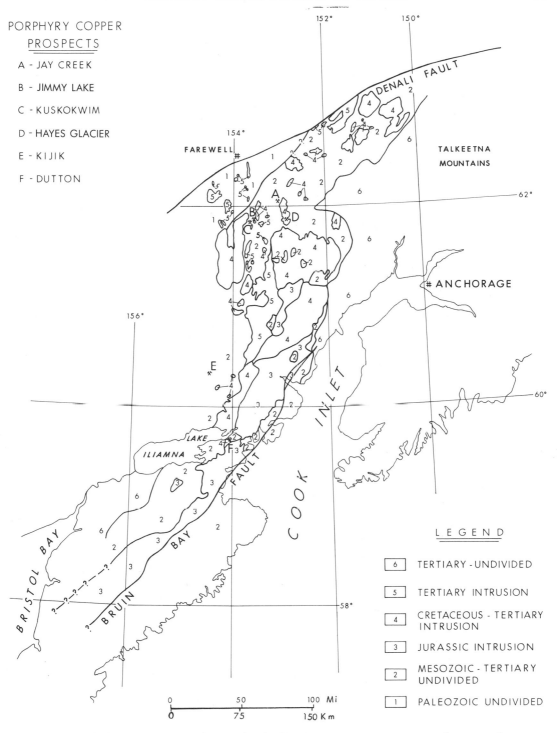

Fig. 29. Geologic index map of Alaskan Peninsula. Porphyry copper prospects are shown as they occur in what is essentially a Jurassic island-arc environment. The porphyry copper occurrences are posttectonic (*modified from Reed and Lanphere, 1972*).

Table 6. Principal Porphyry Copper Prospects of the Continental Margin Belt—Western Section

Name	Quadrangle	Location Lat.	Location Long.	Pluton Type	Pluton Age, m.y.	Pluton Host Rock	Alteration Zoning Sequence From Center	Pyrite Zone Size, m x 10³ (x 10³ ft)	Pyrite Zone % in Phyllic Zone	Structure	Reference
Attu	Attu	52°52';	172°50'	Qtz Dio Por	5.9	Cret Vol	Phy-Arg-Prop	2.4 x 1.8 (8 x 6)	7	Stockwork	Gates, et al., 1971
Unalaska	Unalaska	53°52';	166°32'	Grdr Por	Tert	Tert Vol	Phy-Arg-Prop	1.5 x 1.2 (5 x 4)	3	Stockwork	Berg & Cobb, 1967
Warner Bay	Chignik	56°10';	158°25'	Qtz Dio Por	Tert	Tert Vol	Pot-Phy-Arg-Prop	0.9 x 0.9 (3 x 3)	2	Breccia	Cobb, 1972
Dutton	Iliamna	60°40';	153°56'	Grdr Por	59	Qtz Dio	Phy-Arg-Prop	1.2 x 0.9 (4 x 3)	3	Stockwork	Detterman & Cobb, 1972
Kijik	Lake Clark	60°20';	154°27'	Qtz Mon Por	Tert	Jur Vol	Phy-Arg-Prop	0.9 x 0.9 (3 x 3)	3	Stockwork	Berg & Cobb, 1967
Hayes Glacier	Tyonek	61°35';	152°33'	Qtz Mon Por	Tert	Cret Sed	Phy-Arg-Prop	1.5 x 1.2 (5 x 4)	4	Stockwork	Berg & Cobb, 1967
Jimmy Lake	Lime Hills	61°42';	153°10'	Qtz Mon Por	26(?)	Cret Sed	Phy-Arg-Prop	2.1 x 2.1 (7 x 7)	6	Stockwork	Reed & Elliott, 1970
Jay Creek	McGrath	62°12';	153°45'	Qtz Mon Por	57(?)	Cret Sed	Phy-Arg-Prop	2.1 x 1.5 (7 x 5)	6	Breccia	Reed & Elliott, 1968
Kuskokwim	Lime Hills	61°33';	153°15'	Qtz Dio Por	Tert	Cret Sed	Phy-Arg-Prop	1.5 x 1.2 (5 x 4)	4	Stockwork	Reed & Elliott, 1970
Pyramid Mtn.	Port Moller	55°37';	160°41'	Qtz Dio Por	6.3	Tert Sed	Pot-Phy-Prop	3.7 x 3.0 (12 x 10)	6	Stockwork	Berg & Cobb, 1967
Ivanof	Stepovak Bay	55°53';	159°25'	Qtz Dio Por	Tert	Tert Sed	Pot-Phy-Arg	2.1 x 4.6 (7 x 15)	6	Stockwork	Berg & Cobb, 1967
Dry Creek	Chignik	56°32';	158°23'	Qtz Dio Por	3.3	Jura Sed	Pot-Phy-Prop	2.7 x 2.7 (9 x 9)	4	Stockwork	Berg & Cobb, 1967

Tert: Tertiary
Qtz: Quartz
Dio: Diorite
Por: Porphyry

Cret: Cretaceous
Grdr: Granodiorite
Mon: Monzonite
Vol: Volcanics

Prop: Propylitic
Phy: Phyllic
Arg: Argillic
Jura: Jurassic

closer inspection include porphyry copper deposits. The more important of these deposits are summarized in Table 7.

Most known color anomalies, although not closely investigated, are spatially associated with quartz-bearing Tertiary calc-alkalic intrusions and characteristics of these leached cappings are similar to those outlined in Table 6.

Characteristics of Porphyry Copper Deposits—Table 6 summarizes data thus far known on porphyry copper prospects in the western section of the continental margin belt. Published descriptions generally are old and incomplete and the reference for any particular deposit may provide only skeletal information. The level of information available for deposits included in Table 6 varies and most are raw prospects. Recent exploration by Quintana, Duval, Bear Creek, and Newmont has identified numerous porphyry copper prospects on the Alaska Peninsula other than those listed in Table 6. This volume discusses only those prospects mentioned in the literature and included in Table 6.

Most deposits included in Table 6 are essentially copper-gold type porphyry deposits, although a few (e.g., Kijik, Pyramid, Dutton) may also contain molybdenum within the porphyry center in a zonally developed district. Though copper-gold type porphyry deposits dominate (e.g., Red Mountain: Tyonek Quadrangle, 61°45'. 152°45'), no specific mention of them as such is made in the literature.

Specific examples of copper-molybdenum and copper-gold porphyry copper deposits found in this part of the continental margin belt are cited below.

Pyramid Deposit—Geologic characteristics of the Pyramid porphyry copper deposit are outlined by Armstrong, et al. (1976), whereas the deposit is mentioned by Berg and Cobb (1967) as one of the Pan-American prospects. This description follows Armstrong, et al. (1976).

The Pyramid porphyry copper deposit developed in and near a quartz diorite porphyry stock. The stock intrudes thin remnants of Pliocene(?) andesitic rocks, the widespread

Oligocene Stepovak formation, and the Eocene Tolstoi formation (Burk, 1956) that conformably underlies the Stepovak. Mesozoic trench melange underlies both formations.

The center of mineralization is within the quartz diorite porphyry pluton, a quartz-rich pluton with hornblende and zoned plagioclase phenocrysts. Suites of alteration minerals form well defined zones about the center of mineralization. The biotite-rich potassic core zone consists of dominant secondary biotite, quartz, and plagioclase with lesser magnetite and sulfide. Orthoclase has not been identified. Copper and molybdenum are nearly absent from potassic zone mineralization. Surrounding the potassic zone, a typical phyllic zone consists of felted matlike intergrowths of sericite, quartz, and sulfide. Copper and molybdenum values are restricted entirely to the phyllic zone and gold has not been found to occur significantly in any alteration assemblage. A propylitic zone developed peripherally to the phyllic zone is characterized by chlorite, epidote, and calcite. Fresh rocks outcrop beyond the propylitic zone. Pervasive disseminated pyrite occurs in all zones with the pyrite halo averaging 4000 m in diameter. Hydrothermal biotite has a 6.3 m.y. K-Ar date.

Dry Creek Deposit—Berg and Cobb (1967) mention Dry Creek as one of the Pan-American prospects, but the description by Armstrong, et al. (1976), is followed in this section. The deposit is developed in a quartz diorite stock of porphyritic to equigranular texture that intrudes into the Jurassic Naknek and Stanuikovich formations .(Burk, 1965). Ash flows from recent volcanic eruptions nearby form a thin discontinuous cover over parts of the deposit.

The center of mineralization lies within the quartz diorite stock, which commonly has prominent quartz, zoned plagioclase, hornblende, and biotite phenocrysts. The center coincides with a core potassic alteration silicate assemblage. Peripheral to the potassic zone is a typical phyllic zone that is in turn surrounded by a propylitic zone. Pyrite as a pervasive dissemination exists in all

Table 7. Significant Color Anomalies in Southwestern Alaska

Location		Size, m x 10³ (x 10³ ft) of Pyrite Halo	Pluton (if present)	Structure	% Py	Alteration Zones From Core Outward	Reference if Available
Quadrangle	Coordinates						
Unimak	54°03'; 165°33'	0.9 x 0.9 (3 x 3)	Qtz dio por	Breccia (tour)		Pot-Phy-Arg-Prop	
Umnak	53°20'; 168°22'	1.5 x 0.9 (5 x 3)	Qtz dio por	Stockwork		Phy-Arg-Prop	
Umnak	53°15'; 168°23'	1.5 x 0.3 (5 x 1)	Qtz dio por	Stockwork		Silic-Arg	
Unalaska	53°53'; 166°33'	1.5 x 0.6 (5 x 2)	None	Dissemination		Arg-Prop	Berg & Cobb, 1967
Unalaska	53°45'; 166°10'	1.5 x 0.6 (5 x 2)	None	Dissemination	5%	Arg-Prop	
Unalaska	53°40'; 166°50'	0.9 x 0.5 (3 x 1.5)	Grdr	Stockwork	3%	Phy-Arg-Prop	Berg & Cobb, 1967
Unalaska	53°41'; 166°52	0.6 x 1.2 (2 x 4)	Grdr	Stockwork	3%	Phy-Arg-Prop	Berg & Cobb, 1967
False Pass	54°57'; 163°05'	2.4 x 4.9 (8 x 16)	Dacite (?)	Breccia (tour)		Phy-Arg-Prop	
False Pass	54°52'; 163°17'	1.8 x 2.4 (6 x 8)	Grdr	Breccia (tour)		Arg-Prop	
Port Moller	55°11'; 161°48'	0.9 x 1.8 (3 x 6)	Qtz dio por	Stockwork		Phy-Arg-Prop	
Stepovak Bay	55°49'; 159°54'	0.6 x 1.5 (2 x 5)	Qtz dio	Stockwork		Phy-Prop	
Stepovak Bay	55°48'; 159°33'	1.8 x 3.0 (6 x 10)	Qtz dio	Stockwork		Phy-Arg-Prop	
Chignik	56°11'; 158°58'	1.8 x 3.0 (6 x 10)	Qtz dio	Stockwork		Pot-Prop	
Chignik	56°17'; 159°03'	3.0 x 1.5 (10 x 5)	Qtz dio	Stockwork		Arg-Prop	
Chignik	56°14'; 158°32'	3.0 x 4.6 (10 x 15)	Qtz dio	Stockwork		Phy-Prop	
Chignik	56°13'; 158°42'		Qtz dio	Stockwork		Arg-Prop	

Grdr: Granodiorite
Qtz: Quartz
Pot: Potassic

Phy: Phyllic
Arg: Argillic
Prop: Propylitic

Silic: Silicification
Tour: Tourmaline
Dio: Diorite

zones and the pyrite halo has a diameter of approximately 3000 m. The potassic zone has a 3.3 m.y. K-Ar date.

The potassic zone is characterized by secondary chlorite, biotite, quartz, and plagioclase. Orthoclase has not been identified. The biotite may be altered to chlorite-rutile-calcite-clay mixtures with traces of epidote and apatite. The plagioclase may also be altered. The potassic zone grades outward into a sericite-rich, sulfide-poor phyllic zone containing quartz-sericite veinlets as well as disseminated sericite. A chlorite-rich propylitic zone surrounds the phyllic zone.

Copper occurs as chalcopyrite with magnetite and pyrite disseminated in the potassic zone and in veinlets with quartz and pyrite in the phyllic zone. Molybdenite is not a commonly observed mineral, but parts of the phyllic zone have high gold:copper ratios.

Other Deposits—The Attu porphyry copper is associated with young sodic hornblende quartz-diorite porphyry (called sodic hornblende dacite by Gates, et al., 1971). The granodiorite porphyry mineralized with sulfide on Unalaska, on the other hand, may be part of a zoned calc-alkalic intrusion (Drewes, et al., 1961). Mineralized quartz-diorite porphyry at Warner Bay is variably porphyritic with granitic texture present in part of the stock. The mineralized quartz-monzonite porphyries of the Alaskan Peninsula (e.g., Kijik, Jimmy Lake, Jay Creek, and Hayes Glacier) are all characterized by euhedral quartz phenocrysts.

The change in composition from west to east in the mineralized intrusions, from soda-rich quartz diorite on Attu to quartz monzonite near the Denali fault, coincides with change in the setting of the island arc from oceanic basin to continental margin. In this sense, composition of the mineralized intrusion does seem to reflect crustal environment.

Petrographic details on most porphyry occurrences listed in Tables 6 and 7 are sketchy. However, for those deposits associated with quartz diorite, the pluton is a porphyritic quartz diorite with a microaplitic quartz-dioritic groundmass in most cases. In many cases the granodiorite or quartz monzonite listed in these tables as the intrusive closest to ore spatially and temporally is from plutonic complexes in which the weighted average composition is quartz diorite. Orthoclase may occur as fine-grained interstitial fillings in the groundmass where its origin may be magmatic or metasomatic. Plagioclase, hornblende, and biotite as well as quartz occur as phenocrysts. In general most mineralized intrusives appear to have been forcefully injected and are epizonal.

Alteration zones shown in Tables 6 and 7 indicate a similar response to attack by hydrothermal fluids for all the deposits. Potassic core zones are recognized at Warner Bay, Dry Creek, and Pyramid. Absence of a reported potassic zone in other examples could be due to poor exposures and to lack of exploration in deposits of this part of the belt. However, where the potassic zone has been recognized, the characteristic secondary silicate is biotite, although minor chlorite occurs. Orthoclase may be present but plagioclase is the dominant secondary feldspar occurring with biotite. Where orthoclase is absent, potassium is assumed to be incorporated in the secondary biotite.

Each deposit in Table 6 has a well developed phyllic zone. Quartz-sericite-pyrite zones have variable pyrite contents (from 2% for Warner Bay to 7% for Attu). Sericite and quartz dominate as fine-grained intergrowths in this zone, although biotite, calcite, plagioclase, pyrophyllite, and chlorite have also been identified. Copper is deposited most significantly in the sericite zone of most deposits but occurs importantly in the potassic zone of Warner Bay and Dry Creek.

An argillic zone containing weak pyrite, clay minerals, and carbonate peripheral to the phyllic zone is identifiable in most deposits. It is smaller than the phyllic zone and is usually free of significant hypogene copper mineralization.

Peripheral to the argillic zone is the usual chlorite-rich, pyrite-poor propylitic zone. Zinc, lead, or precious metals appearing in or close to a porphyry deposit (e.g., Kijik,

Jay Creek, or Jimmy Lake) tend to occur in the propylitic zone.

Pyrite in each deposit oxidizes rapidly to form large easily distinguished halos. Oxidation of pervasive pyrite can be seen even in areas of recent glaciation, and color anomalies developed within these deposits constitute a handy prospecting tool in the search for others. Pyrite-caused color anomalies are known erratically between known deposits from Attu eastward through the Aleutian chain to the Denali fault. Eventually these will be evaluated and some may be found to carry gold-copper type deposits (Table 7).

Tourmaline occurs sporadically in porphyry prospects east from Unalaska to Cook Inlet, although it is not reported as a prominent mineral in pervasive alteration zones west of this island. Nowhere, however, does it appear to be a significant alteration mineral.

Radiometric dates on porphyry copper prospects are rare in this belt. The Jimmy Lake stock has not been dated, but Reed and Lanphere (1972) provide a K-Ar date of 26.8 m.y. for the larger quartz monzonite pluton associated with it. Field relationships suggest a similar age for the mineralized intrusion, so this date is used in Table 6. Jay Creek mineralization is near a stock with a 57.0 m.y. K-Ar date (Reed and Lanphere, 1972). Their age probably represents the age of mineralization, so this is used in Table 6. The youngest significantly metallized porphyry copper deposit in Alaska is Pyramid, for which Armstrong, et al. (1976), provide a 6.3 m.y. K-Ar date on hydrothermal biotite. Armstrong, et al. (1976), also provide a 3.3 m.y. K-Ar date on secondary silicates intergrown with chalcopyrite from Dry Creek. Hayes Glacier mineralization appears to cut a stock with a K-Ar date of 69.7 m.y. (Reed and Lanphere, 1972). That date is not used in Table 6 inasmuch as other dates have been found in this area on igneous rocks, both younger and older. The Kijik porphyry copper deposit (also known as the Thompson prospect) lies west of the Merrill Pass quartz monzonite. The Merrill Pass pluton has a K-Ar date of 38.4 m.y. near Kijik (Reed and Lanphere, 1972). No

date is available on the Kijik deposit itself, but the 38.4 m.y. date on an intrusion with similar composition suggests that Kijik is Tertiary also. A preore intrusion at the Dutton deposit was found to have a K-Ar date of 59.6 m.y. (Reed and Lanphere, 1969). Mineralization followed intrusion, and it could therefore be 59 m.y.

In summary, the more important porphyry copper occurrences thus far discovered in the Aleutian arc and in the western section of the continental margin belt are associated with calc-alkalic plutons in a geological environment that suggests that intrusion and mineralization were largely accomplished within the last 60 m.y. Older deposits may be found on the Alaska Peninsula close to the Denali fault; however, it is surmised from stratigraphic studies that deposits with ages less than 40 m.y. more commonly developed.

Structural elements in individual porphyry copper deposits do not seem to reflect the underthrusting associated with subduction as proposed by Pitman and Hayes (1968). On the other hand, plate motions of Clague and Jarrard (1973) have not been utilized in the past to explain structural details found in individual deposits. Strike-slip faulting with unknown amounts of displacement have attended development of each deposit in Tables 6 and 7, and in each a major stockwork set is oriented parallel to the axis of the Aleutian Islands or the Peninsula, as the case may be. Neither the involvement of strike-slip faults in these deposits near the plate boundary, the fracture pattern in each deposit, nor the age of deposits conflicts with broad-scale plate motions proposed by Cormier (1975) and Clague and Jarrard (1973).

Eastern Section

Introduction: The eastern section of the continental margin belt has the greatest variety of porphyry copper types and the largest concentration of prospects in Alaska (Hollister, et al., 1975). It extends east and southeast from Cook Inlet to the southern tip of the Alaskan panhandle. Like the western section, it is limited on the north by the Denali fault. The Coast Range plutonic complex forms the eastern boundary of this belt; this zone overlaps parts of the Yukon and British Columbia as well as Alaska.

Examples of porphyry copper deposits cited in this chapter are representative of deposits in this area. It is not the intention to list all known porphyry occurrences; rather, enough are mentioned to show the character of the deposits.

Geology: The eastern section of the continental margin belt is dominantly a province of allochthonous terranes of Paleozoic age and younger (Fig. 30). Included in this province is an allochthonous fragment of rocks of continental affinity underlying much of southeastern Alaska (the Alexander Terrane of Berg, et al., 1972). Rocks ranging in age from pre-Ordovician to Upper Triassic are found in this allochthon. The terrane, a heterogeneous assemblage of volcanic, sedimentary, and metamorphic rocks, includes granitic plutons of Silurian age. Rocks as young as Upper Triassic in the Alexander allochthon suggest that displacement of the fragment could have continued then.

With the exception of Alexander terrane, most rocks south of the Denali fault represent a series of sedimentary and volcanic deposits continuously accreting to the continent since late Paleozoic time. The oldest strata are mostly submarine flows, breccias, and pyroclastic rocks interlayered with volcanoclastic rocks of Pennsylvanian-Permian age that are considered part of an extensive andesitic volcanic system. The arc assemblage was built directly on oceanic crust (Richter and Jones, 1973) and appears similar in position to the pre-Mesozoic volcanics of Vancouver Island, which Stacey (1973) interprets as lying directly on oceanic crust. Deposited on the remnants of this arc sequence are younger Permian marine sedimentary rocks—a large volume of Triassic basalts, minor Triassic marine limestone, and a wide diversity of Mesozoic rocks.

Subsequent to deposition of the Triassic rocks, three arc-trench systems (Lower Jurassic, a continuation of the system well developed in the western section; Upper Jurassic-Lower Cretaceous, also described as the Gravina-Nutzotin belt of Berg, et al., 1972; and a Tertiary assemblage) have added material to the continental margin and contiguous oceanic trench. Rocks of the Lower Jurassic arc-trench system are best exposed west of the Talkeetna Mountains in the western part of the eastern section. The Upper Jurassic-Lower Cretaceous system is most prominent in the southeastern Alaskan panhandle in a belt parallel to coeval parts of the Coast Range plutonic complex (Berg, et al., 1972). Because Lower Jurassic arc-trench rocks are minimally exposed in this area, Upper Jurassic-Lower Cretaceous rocks may rest directly on the allochthonous Alexander terrane. This relationship implies that tectonic displacement of that allochthon could have continued into Lower Jurassic.

Radiometric dating of plutonic rocks in the eastern part of the continental margin province indicates five principal periods of intrusive activity: 285-282 m.y., 179-154 m.y., 117-105 m.y., 83-58 m.y., and 41-25 m.y. (Lanphere and Reed, 1973). Granitic rocks of the three oldest periods are coeval with and spatially related to the three arc-trench systems, possibly representing roots of the volcano-plutonic belts of these systems. Plutons of the youngest period (Tertiary) of intrusive activity do not appear associated with known deposits of Tertiary volcanogenic rocks and hence may not be arc-trench derivatives.

Porphyry copper deposits of the eastern continental margin appear to be spatially related to plutons associated with all the known arc-trench systems. Regardless of alternate hypotheses that may be proposed to explain the presence of widespread marine volcanic rocks and the Alexander allochthon, porphyry copper deposits penetrate a wide variety of crust types in the southeastern part of the state.

In contrast to deposits of other belts, both quartz-rich and quartz-deficient intrusions host porphyry copper deposits in this province. However, the principal deposits occur in silica-rich rocks generally associated with Cretaceous plutons.

Characteristics of Porphyry Copper De-

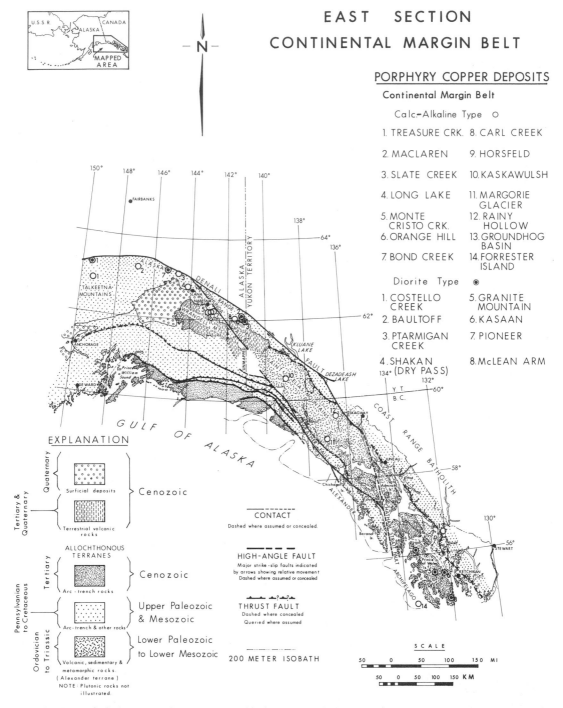

Fig. 30. Geologic index map of southeastern Alaska. Some of the more important porphyry prospects of southern and southeastern Alaska are shown in their geologic setting. Porphyry copper deposits associated with quartz-deficient intrusions—diorite type, are distinguished from deposits with quartz-rich plutons—calc-alkalic type (*modified from Berg, et al., 1974*).

posits: Porphyry copper occurrences in the eastern section of the continental margin belt separate into two lithologic types based on petrography of the host intrusion. The largest number of occurrences are associated with quartz-bearing calc-alkalic intrusions resembling the porphyry copper deposits compiled by Lowell and Guilbert (1970). In this province deposits have been referred to as the calc-alkalic type rather than the Lowell and Guilbert (1970) model because of some persistent peculiarities. It would simplify nomenclature, however, if the Lowell and Guilbert (1970) model were modified to include these quartz-bearing calc-alkalic mineralized plutons. The calc-alkalic type is summarized in Table 9. The other type of occurrence arbitrarily referred to as the diorite type (Hollister, et al., 1975) includes alkalic quartz-free mineralized intrusions as well as dioritic calc-alkalic mineralized plutons. Field distinction between alkalic and calc-alkalic plutons may be difficult due to alteration. For this reason, the term diorite is preferred for all porphyry copper deposits associated with plutons lacking quartz as a magmatic mineral. Use of the term diorite model lessens the need for field separation into alkalic and calc-alkalic quartz-deficient types when positive identification may be difficult. Deposits of this type are summarized in Table 8. Table 10 compares general characteristics of the two types.

Most deposits listed in Tables 8 and 9 have benefited from a pilot exploration program. Further exploration may modify data summarized in the tables, but such modifications probably will be inconsequential. References from the literature are cited in both tables, but in each the general geologic setting is described while details of the porphyry copper deposits are largely omitted. Details of the deposits are from industry sources. Names given in the two tables are from the literature. However, should the names be unfamiliar to those working in the area, identity of the deposit may still be established by latitude and longitude.

Porphyry copper deposits containing higher concentrations of copper and molybdenum tend to occur in two distinct areas, one extending from Long Lake deposit southeasterly to the Horsfeld, the other extending from Kasaan south to McLean Arm. The northernmost group is composed mostly of calc-alkalic type deposits.

Diorite-Type Deposits—Diorite-type deposits are associated with mineralized quartz-free intrusive and extrusive rocks. The model was proposed originally to facilitate economic evaluation of quartz-deficient and nonsericite-bearing porphyry copper plutons and remains a field exploration tool with only local use in specific terranes. Individual deposits of this type from Alaska have not been described in detail in the past, so much data in this chapter are new to the literature. The most common host rocks are diorite, syenite, or monzonite (Hollister, et al., 1975). One may speculate that both alkalic and calc-alkalic quartz-deficient porphyry copper igneous hosts react similarly to metasomatising hydrothermal fluids regardless of petrogenic origin. Both have greater amounts of iron, alkali earths, and metals; hence their final response to alteration reflects the surplus of these constituents over quartz-phenocryst-bearing calc-alkalic hosts. Hydrothermal fluids that attack diorite-type hosts ordinarily do not contain enough sulfur to consume all the iron as pyrite. Metasomatic addition of potassium and/or sodium in the mineralization center of the diorite-type deposits thus promotes occurrence of secondary biotite with metallic sulfides; secondary albite commonly occurs with biotite. In some deposits orthoclase is also present. Chlorite may occur with the biotite as well. Other iron-bearing silicates (e.g., chlorite and epidote) tend to occur outside of and overlapping onto the biotite zone, but sericite occurs pervasively only rarely.

The potassic-phyllic-argillic-propylitic zoning of the Lowell and Guilbert (1970) model ordinarily fails to develop in dioritic hosts. In terms of the mineralogic description given alteration zones by Lowell and Guilbert (1970), it is possible in diorite porphyries to proceed directly from the central potassic to the outer propylitic zone without encounter-

Table 8. Principal Diorite Model Porphyry Copper Prospects of the Continental Margin—Eastern Section

Name	Location			Pluton			Alteration Zoning Sequence From Center	Pyrite Zone		Structure	Reference
	Quadrangle	Lat.	Long.	Type	Age, m.y.	Host Rock		Size, m x 10^3 (x 10^3 ft)	Content in Copper Zone, %		
Costello Cr.	Healy	63°15';	149°28'	Diorite	Cret	Cret Sed	Prop	0.9 x 1.2 (3 x 4)	2	Stockwork	Hawley, et al., 1969
Baultoff	Nabesna	62°06';	141°14'	Diorite	114	Cret Sed	Pot-Prop	0.9 x 0.9 (3 x 3)	3	Stockwork	Richter, et al., 1974
Ptarmigan Creek	McCarthy	61°58';	141°03'	Diorite	114	Cret Sed	Pot-Prop	1.5 x 0.3 (5 x 1)	1	Stockwork	Moffit & Knopf, 1910
Shakan (Dry Pass)	Petersburg	56°08';	133°25'	Diorite	?	L. Paleo Sed	Prop	0.9 x 0.9 (3 x 3)	2	Breccia	Herreid & Kaufman, 1964
Granite Mtn.	Craig	55°32';	132°40'	Diorite	?	Paleo Vol	Prop	0.9 x 0.9 (3 x 3)	1	Stockwork	Berg & Cobb, 1967
Kasaan	Craig	55°31';	132°18'	Diorite	?	Paleo Vol	Prop	0.9 x 1.2 (3 x 4)	2	Stockwork	Warner, et al., 1961
McLean Arm	Dixon Entrance	54°48';	130°14'	Diorite	Trias?	Dev Vol	Prop	0.6 x 1.5 (2 x 5)	1	Stockwork	

Cret: Cretaceous   Trias: Triassic   Prop: Propylitic
Paleo: Paleozoic   Jur: Jurassic   Pot: Potassic
Vol: Volcanic   Sed: Sediment

Table 9. Principal Quartz Monzonite-Quartz Diorite Type Porphyry Copper Prospects of the Continental Margin—Eastern Section

Name	Location			Pluton			Alteration Zone Sequence From Center	Pyrite Zone		Structure	Reference
	Quadrangle	Lat.	Long.	Type	Age, m.y.	Host Rock		Size, m x 10^3 (x 10^3 ft)	Content (%) in Phyllic Zone		
Treasure Creek	Talkeetna Mtn.	62°52';	149°20'	Qtz Mon Por	Tert	Cret Sed	Phy-Arg-Prop	1.2 x 1.2 (4 x 4)	3	Stockwork	Hollister, et al., 1975
Maclaren	Mt. Hayes	63°15';	146°45'	Qtz Dio Por	Tert	Trias Vol	Pot-Phy-Arg-Prop	1.2 x 3.7 (4 x 12)	6	Stockwork	Smith, 1971; Kaufman, 1967
Slate Creek	Mt. Hayes	63°09';	144°52'	Qtz Dio Por	?	Perm Vol	Pot-Arg-Prop	1.2 x 1.8 (4 x 6)	4	Stockwork	Rose, 1967
Long Lake	Gulkana	62°49';	144°10'	Qtz Mon Por	282	Penn Vol	Pot-Phy-Prop	0.9 x 1.5 (3 x 5)	3	Stockwork	Richter, 1966
Monte Cristo Cr.	Nabesna	62°12';	143°00'	Grdr Por	109	Cret Vol	Arg-Prop	0.9 x 1.2 (3 x 4)	3	Stockwork	Richter, 1974
Orange Hill	Nabesna	62°12';	142°51'	Qtz Dio Por	105	Perm Vol	Pot-Phy-Arg-Prop	1.8 x 3.0 (6 x 10)	4	Stockwork	Richter, 1974
Bond Creek	Nabesna	62°13';	142°44'	Grdr Por	109	Perm Vol	Pot-Phy-Arg-Prop	2.7 x 4.9 (9 x 16)	5	Breccia	Richter, 1974
Carl Creek	Nabesna	62°03';	141°36'	Qtz Mon Por	111	Perm Vol	Phy-Arg-Prop	0.9 x 0.9 (3 x 3)	4	Stockwork	Richter & Jones, 1973
Horsfeld	Nabesna	62°03';	141°14'	Qtz Mon Por	111	Cret Sed	Arg-Prop	0.9 x 0.9 (3 x 3)	3	Breccia	Richter, et al., 1974
Kaskawulsh	Mt. St. Elias	66°33';	139°00'	Qtz Dio Por	Tert	Perm Sed	Pot-Prop	0.6 x 0.9 (2 x 3)	2	Stockwork	Hollister, et al., 1975
Margorie	Skagway	59°02';	137°05'	Grdr	Tert	Paleo Meta	Phy-Arg-Prop	0.9 x 1.2 (3 x 4)	3	Stockwork	MacKevett, et al., 1971
Rainy Hollow	Tatshenshini	59°35';	136°35'	Qtz Dio Por	Tert	Perm Sed	Pot-Phy-Arg-Prop	1.2 x 1.2 (4 x 4)	3	Breccia	Hollister, et al., 1975
Gr. Hog Basin	Petersburg	56°31';	132°06'	Granite	Tert	Paleo Meta	Pot-Phy-Prop	1.2 x 1.2 (4 x 4)	4	Stockwork	Hollister, et al., 1975
Forrester Is.	Dixon Ent.	54°50';	133°30'	Qtz Mon Por	Tert	Cret Sed	Phy-Arg-Prop	?	4	Stockwork	Clark, et al., 1971

Cret: Cretaceous   Penn: Pennsylvanian   Qtz: Quartz   Phy: Phyllic   Grdr: Granodiorite
Tert: Tertiary   Paleo: Paleozoic   Dio: Diorite   Por: Porphyry   Arg: Argillic
Perm: Permian   Meta: Metamorphics   Prop: Propylitic   Pot: Potassic

Table 10. Comparison of Characteristics of the Quartz Monzonite-Quartz Diorite and the Diorite Models

Feature	Quartz Monzonite—Quartz Diorite Type	Diorite Type (if Sodic Metasomatism Absent)
Intrusive Relationships		
Typical Intrusion Close to Ore	Quartz Monzonite Porphyry	Syenite Porphyry
Other Intrusions Usually Present	Quartz Diorite Porphyry	Diorite Porphyry
Alteration		
Central Core Area	Potassic Zone (including Orthoclase-Biotite and/or Orthoclase-Chlorite)	Potassic Zone (including Orthoclase-Biotite and/or Orthoclase-Chlorite)
Peripheral to Core	Phyllic Zone (Quartz-Sericite-Pyrite)	Propylitic Zone (Chlorite-Epidote)
Peripheral to Phyllic	Argillic Zone	—
Peripheral to Argillic	Propylitic Zone (Chlorite-Epidote)	
Where Pervasive Pyrite Occurs	Includes Potassic, Phyllic Zones	Usually Potassic Zone Only
Mineralization		
Quartz in Fractures	Common	Erratic
Orthoclase in Fractures	Common	Erratic
Albite in Fractures	Trace	Erratic
Magnetite	Minor	Common
Pyrite in Fractures	Common	Common
Molybdenite	Common	Rare
Chalcopyrite:Bornite Ratio	3 or greater	2 or Less
Dissemination of Chalcopyrite	Present	Important
Gold	Rare	Important
Structure		
Breccia	May Occur	Very Rare
Stockwork	Important	Important

ing phyllic or argillic zones. Hypogene copper may be found in both zones. These deposits commonly have abnormal gold:copper ratios but subnormal molybdenum:copper ratios.

Most diorite porphyry copper deposits are associated with pre-Upper Cretaceous rocks (Hollister, et al., 1975). These deposits are less likely to be posttectonic; their distribution is shown in Fig. 30.

Baultoff exemplifies a diorite-model porphyry copper deposit where mineralization occurs in a calc-alkalic diorite pluton. A small quartz-bearing porphyry plug exists in the mineralized zone, which has a quartz-rich aplitic groundmass, suggesting a differentiation sequence from diorite to quartz diorite. Mineralization consists of weak chalcopyrite dissemination associated with secondary biotite in a potassic zone. Chalcopyrite-pyrite veinlets cut chloritized diorite in a propylitic zone adjacent to the biotized diorite. No sericite (phyllic) zone has developed and hypogene copper sulfides occur in both the potassic (biotite) and the propylitic (chlorite) zones adjacent to it. Other diorite examples are listed in Fig. 30.

Calc-Alkalic-Type Deposits—Calc-alkalic-type deposits include quartz diorite, granodiorite, and quartz monzonite plutons affected by porphyry copper-type alteration and mineralization. Alteration zones in this type are classified like those compiled by Lowell and Guilbert (1970). The feature distinguishing them from the Lowell and Guilbert (1970) model as it is now used is the absence or deficiency of secondary orthoclase in the Alaskan calc-alkalic potassic zones. Secondary plagioclase occurs with biotite in the potassic zone generally in amounts exceeding those of orthoclase. The unifying characteristic of this model is frequent presence of sericite in a phyllic zone. The 14 examples of quartz-bearing intrusions hosting porphyry copper deposits selected for inclusion in Table 9 are believed typical of the calc-alkalic model for the eastern section of the continental margin belt.

Petrography—No consistent relationship exists between composition of the mineralized intrusion and type of crust penetrated for deposits in the eastern section of the continental margin belt. Long Lake exemplifies a mineralized quartz monzonite that intrudes Upper Paleozoic volcanic rocks. Slate Creek, Orange Hill, and Bond Creek are examples of mineralized quartz diorite intruding into the same unit. The porphyry copper deposit at Kaskawulsh is a quartz diorite porphyry that could have penetrated a thick subsurface section of marine volcanic rocks, making its composition and setting consistent with Orange Hill. The quartz monzonite porphyry on Forrester Island, on the other hand, differs in crustal setting from Long Lake since it intrudes a Cretaceous trench sediment accumulation. Lack of an intrusion composition-crust type correlation with quartz-bearing mineralized intrusions, and particular difficulty in matching intrusive composition with wall-rock environments, is more accentuated in this portion of Alaska than elsewhere in the Cordilleran orogen. Most plutons associated with porphyry deposits are epizonal forcefully emplaced stocks.

As noted in Table 9, mineralized intrusions may range in composition from quartz diorite to quartz monzonite for the phase spatially and temporally closest to ore. Most mineralized rocks are parts of plutonic complexes that include quartz diorite even if quartz diorite is not the mineralized phase. Quartz diorite mineralized plutons (e.g., Orange Hill, Bond Creek) have no orthoclase phenocrysts and their groundmasses are quartz-rich and aplitic. Characteristic $Na_2O:K_2O$ ratios for mineralized plutons of the continental margin belt are highly variable from one pluton to the next.

ALTERATION. All examples listed in Table 9 have alteration zones predictable by the Lowell and Guilbert (1970) model, although some zones commonly are missing. Most well investigated porphyries in this section have a secondary biotite- or chlorite-dominant zone in their core that may contain some secondary orthoclase. Secondary plagioclase is characteristic in all deposits, however. Where the original intrusion has a high $Na_2O:K_2O$ ratio, the core tends to be impoverished in orthoclase. Erratic presence of orthoclase with a biotite-chlorite zone suggests that this zone should be included with the potassic zone of the Lowell and Guilbert (1970) model even if the mineralogy is distinguished by an absence of orthoclase in some deposits. Megascopically a potassic zone devoid of orthoclase but still containing abundant biotite may be indistinguishable from a potassic zone rich in K-feldspar.

In most cases the copper-molybdenum-gold content of the potassic zone is lower than that of the phyllic zone surrounding it. For example, the potassic zone of the MacLaren deposit is virtually barren and at Bond Creek copper sulfide occurs in this zone only in minor disseminations and as vein and breccia filling.

In most deposits shown in Fig. 29, typical pyrite-sericite-quartz-rich phyllic zones adjoin the potassic zones. At Bond Creek the phyllic zone is rich in quartz and stands out as a resistant topographic high, whereas some phyllic zones (e.g., Treasure Creek) contain little silica and form topographic lows. Pyrite content in the phyllic zone is highly variable, ranging from 3% to 6%. Copper and molybdenite occur most importantly in the phyllic zone but metallic concentrations are erratic.

An argillic zone may be adjacent to the phyllic zone. In general, the larger the pyrite halo around the potassic zone the more likely the argillic zone is to develop. Characteristic mineralogy includes clay minerals and pyrite. Copper has not been found in significant amounts in the argillic zone of those deposits explored. Lead and zinc sulfides may begin to occur importantly in this zone, but as isolated vein deposits.

A typical chlorite-rich propylitic zone surrounds the other alteration zones. Calcite, epidote, albite, and erratic pyrite also commonly occur in this zone.

Mineralization appears to be either the copper-molybdenum or the copper-gold type. A correlation exist between low $Na_2O:K_2O$ ratios and high molybdenum:copper ratios for some deposits (e.g., Forrester and Mar-

gorie). Those deposits with high $Na_2O:K_2O$ tend to carry higher gold:copper ratios (Orange Hill, Maclaren, Bond Creek). Both copper-gold and copper-molybdenum deposits may be associated with zoned lead-zinc-copper districts having porphyry copper cores.

The oldest deposit thus far dated in the Cordilleran orogen is the Long Lake mineralized quartz monzonite, with a 282 m.y. K-Ar date (Hollister, et al., 1975). It intrudes coeval Pennsylvanian volcanics and is apparently anomalous in that it is a potash-rich, quartz-bearing intrusion invading marine andesites. Coeval and compositionally similar potassium-feldspar-rich rocks also intrude the Alexander terrane.

The most important deposits, Bond Creek and Orange Hill, have K-Ar dates of 109 and 105 m.y., respectively. Close proximity permits some overlap of their pyrite halos and their geologic setting at the margins of the small Nabesna batholith suggests a similar origin. Carl Creek and Horsfeld (Hollister, et al., 1975) each have a K-Ar date of 111 m.y. Fig. 31 gives the general geology around these dated deposits and includes other porphyry prospects not included in Table 9 or Fig. 20. No other firm dating has been established for all deposits shown in Table 9, although mineralized intrusions are known to have Tertiary dates (e.g., Willow Creek, 70 m.y., Hollister, et al., 1975) and Triassic dates (e.g., Texas Creek, 205 m.y., Hollister, et la., 1975).

Orange Hill and Bond Creek are briefly summarized in this section as examples of quartz-rich calc-alkalic-type porphyry copper deposits.

ORANGE HILL, NABESNA QUADRANGLE. As tabulated by Hollister, et al. (1975), the Orange Hill deposit contains a poorly developed potassic zone, a highly silicic phyllic zone that grades into an erratic argillic zone, and a very extensive propylitic zone. The deposit is centered about a quartz diorite porphyry near where a larger granitic-textured quartz diorite batholith (the Nabesna pluton) makes contact with Upper Paleozoic marine volcanic rocks.

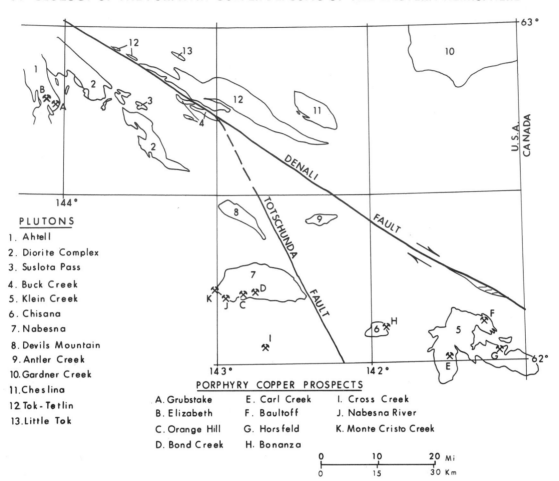

PLUTONS

1. Ahtell
2. Diorite Complex
3. Suslota Pass
4. Buck Creek
5. Klein Creek
6. Chisana
7. Nabesna
8. Devils Mountain
9. Antler Creek
10. Gardner Creek
11. Cheslina
12. Tok - Tetlin
13. Little Tok

PORPHYRY COPPER PROSPECTS

A. Grubstake   E. Carl Creek   I. Cross Creek
B. Elizabeth   F. Baultoff   J. Nabesna River
C. Orange Hill   G. Horsfeld   K. Monte Cristo Creek
D. Bond Creek   H. Bonanza

Fig. 31. Porphyry copper deposits of the Nabesna area. The Nabesna area contains the most intensively explored, most numerous, and best known porphyry copper deposits in Alaska. Most deposits occur on the margins of Cretaceous batholiths and are associated with plutons having quartz phenocrysts. Included in this figure are prospects not mentioned in the text nor included on the tables (*modified from various US Geological Survey maps*).

The potassic zone has a dominant secondary chlorite-biotite-plagioclase mineralogy with no orthoclase. Absence of orthoclase as a secondary silicate in this zone is typical of other quartz-bearing porphyry deposits nearby. Vein quartz and silicification dominate the phyllic zone, although sericite and pyrite still characterize this zone. A kaolin-bearing argillic zone is poorly exposed adjacent to and near this assemblage. Pyrite occurs pervasively with clay along with minor sericite and chlorite. The propylitic zone is characterized by secondary chlorite, calcite, sodic plagioclase, epidote, and minor clay.

Orange Hill has the largest potential reserve of any porphyry copper deposit in this belt and therefore has had more careful screening than other deposits nearby. The strongest molybdenum:copper ratio occurs in the phyllic zone, although molybdenite has been found in both potassic and argillic zones. Hypogene copper sulfide occurs in the potassic, phyllic, and propylitic zones and in skarn deposits in the intruded rock. It is found mostly in veinlets with quartz and other sulfides. Hollister, et al. (1975), report a K-Ar date of 105 m.y. for the occurrence.

BOND CREEK, NABESNA QUADRANGLE. As outlined by Hollister, et al. (1975), the Bond Creek deposit has a potassic zone surrounded by a phyllic zone. External to the phyllic zone are argillic and large propylitic zones. The pyrite halo around the ore deposit includes two centers of mineralization, one dated at 109 m.y. and the other at 24 m.y. Only the older center has potassic, phyllic, and argillic alteration assemblages and it alone is associated with significant copper deposition.

The potassic zone contains secondary biotite, chlorite, quartz, and sodic plagioclase, but no orthoclase. It relates spatially to an irregular leucocratic quartz diorite porphyry intruded near the contact between a granitic-textured quartz diorite and Upper Paleozoic marine volcanic rocks. Quartz, sericite, and pyrite characterize the phyllic zone and kaolin appears dominant in the argillic zone. The propylitic zone contains abundant secondary chlorite, plagioclase, epidote, and minor pyrite.

The most widespread hypogene copper occurs in a stockwork in the potassic and phyllic zones of the pluton near its contact or in skarn in the intruded volcanic rocks. Large breccia pipes occur northeast of the area of strongest copper mineralization but apparently these are not strongly metallized. Molybdenum occurs weakly and erratically in the phyllic zone. Anomalously high gold:copper ratios have been recorded in the skarn.

Comparison of the Two Types—Reconnaissance study of the eastern section of the continental margin belt reveals significant economic and mineralogic differences between what are described as diorite and calc-alkalic types of porphyry copper deposits. Table 10 shows some distinctions between these deposits. The diorite types are more commonly copper-gold porphyry deposits, while the calc-alkalic include copper-molybdenum porphyries.

The original andesitic, dioritic, diabasic, monzonitic, or syenitic composition of the mineralized diorite-model host rocks plays an important part in ultimate characteristics of the alteration zones. Even if hydrothermal fluids involved in the two models were grossly similar (and distinct metallization for each suggests that this was not the case), chemical differences in the two types of intrusive hosts would lead to different reaction products. Where rocks are invaded by hydrothermal fluids with high sulfur potential, pyrite, quartz, and sericite are abundantly developed. The absence of sericite and typical phyllic-zone alteration minerals in the diorite-type deposits suggests that mineralizing fluids associated with these deposits had a low sulfur potential. Most diorite types have smaller pyrite halos with a pyrite zone comparatively diminished in both size and amount. Magnetite is ubiquitous in diorite-model porphyry deposits. On the other hand, less abundant iron was largely converted to pyrite in the quartz-rich hosts while anhydrous as well as hydrous silicates developed in the potassic and phyllic zones.

The resultant distribution of alteration halos for the quartz-bearing calc-alkalic type, where metasomatism has been noted, is a potassic core with successive cones or cylinders of phyllic, argillic, and propylitic alteration. The resultant distribution of alteration halos for the diorite type, if potassic metasomatism was significant, is core potassic with propylitic alteration peripheral to that. Where sodium metasomatism occurred rather than potassic, a typical orthoclase-bearing potassic zone may not have formed.

Mineralization in each model tends to occur as fracture filling in the host intrusion as well as a dissemination. Fractures containing hypogene copper sulfide in the diorite model may not be accompanied by quartz. Typically, the gangue in a diorite type is dolomite, hydrated iron-bearing silicates, and pyrite with quartz in a diminished role. Gangue minerals in the fractures of calc-alkalic type deposits include quartz, pyrite, potassic, or phyllic zone silicates with copper sulfide. Quartz commonly is the most prominent of these minerals.

CONCLUSIONS

Continental margin deposits in Alaska possess characteristics distinct from those found in deposits formed in a cratonic setting. Intrusions associated with porphyry copper deposits vary broadly in the continental margin, ranging from alkalic diorites to calc-alkalic quartz monzonites. The wide compositional range of host rock has provided a variety in type and amount of alteration. Metals emplaced in a continental margin deposit may be correlated occasionally with crustal setting. The crust strongly influenced metal deposition within a mineralized intrusion in porphyry belts interior from the continental margin. Where stocks cut earliest Cretaceous marine volcanic rocks in the Hogatza plutonic belt they may contain copper. Where Precambrian rocks are the host in this belt, mineralized intrusions are nearly devoid of copper. The relationship porphyry copper deposits have to marine volcanic rocks in the Hogatza belt implies that at least some copper in these deposits was acquired from the invaded terrane.

On the other hand, partial independence of petrography and metallogeny from rock type within the continental margin crust suggests at least some metallogenic independence of the porphyry deposit from intruded terrane.

What is called the calc-alkalic type in the continental margin areas of Alaska may be incorporated into the Lowell and Guilbert (1970) model if secondary plagioclase may be admitted as a prominent or unique metasomatic feldspar in the potassic zone. Phyllic, argillic, and propylitic zones of the Alaskan calc-alkalic model are typical of the Lowell and Guilbert (1970) model, as is sulfide ore mineralogy and the structure of the deposit. Those deposits with high sodium:potassium ratios (e.g., Orange Hill and Bond Creek) are distinct in the nature of secondary feldspars. It would seem logical to expand the Lowell and Guilbert (1970) model to include those calc-alkalic porphyry copper deposits with high sodium:potassium ratios if the alteration zones (biotite, phyllic, argillic, and propylitic) are present and if the mineralization is typically the porphyry copper type. Copper occurs most importantly in the phyllic zone but may also be found in the potassic zone in this type of deposit.

A suite of porphyry copper deposits that lacks phyllic zones, is relatively rich in gold, lean in molybdenum, and generally has quartz-free hosts, does not conform to the Lowell and Guilbert (1970) model. These are called diorite-model porphyry copper deposits and copper may occur in the potassic and propylitic zones in such deposits.

REFERENCES AND BIBLIOGRAPHY

"Arctic Geology," 1973, M. G. Pitcher, ed., Memoir No. 19, American Association of Petroleum Geologists, pp. 408-420.

Armstrong, R. L., Harakal, J. E., and Hollister, V. F., 1976, "Late Cenozoic Porphyry Copper," *Transactions,* Institution of Mining and Metallurgy, Vol. 85.

Berg, H. C., and Cobb, E. H., 1967, "Metalliferous Lode Deposits of Alaska," Bulletin No. 1246, US Geological Survey, 245 pp.

Berg, H. C., Jones, D. L., and Richter, D. H., 1972, "Gravina-Nutzotin Belt: Tectonic Significance of an Upper Mesozoic Sedimentary and Volcanic Sequence in Southern and Southeastern Alaska," Professional Paper No. 800-D, US Geological Survey, pp. D1-D24.

Brosge, W. P., and Reiser, H. N., 1964, "Geologic Map and Section of the Chandalar Quad., Alaska," Miscellaneous Geologic Map No. 1-375, US Geological Survey.

Brosge, W. P., et al., 1969, "Probable Permian Age of the Rampart Group, Central Alaska," Bulletin No. 1294-B, US Geological Survey.

Burk, C. A., 1965, "Geology of the Alaska Peninsula—Island Arc and Continental Margin," Memoir No. 99, Geological Society of America, 250 pp.

Clague, D.A., and Jarrard, R. D., 1973, "Hot Spots and Pacific Plate Motion," Transactions, American Geophysical Union, Vol. 54, No. 4, p. 238.

Clark, A.L., et al., 1971, "Reconnaissance Geology and Geochemistry of Forrester Island National Wildlife Refuge, Alaska," Open File Report, US Geological Survey.

Clark, A.L., et al., 1974, "Metal Provinces of Alaska," Map No. 1-834, US Geological Survey.

Clark, K. F., 1972, "Stockwork Molybdenum Deposits in the Western Cordillera," Economic Geology, Vol. 67, pp. 713-758.

Cobb, E. H., 1967, "Chignik," Miscellaneous Map No. MF-37, US Geological Survey.

Cormier, V. F., 1974, "Tectonics of the Aleutian Arc," Bulletin, Geological Society of America, Vol. 86, pp. 443-453.

Detterman, R. L., and Cobb, E. H., 1972, "Iliamna," Miscellaneous Map No. MF-364, US Geological Survey.

Drewes, H., et al., 1971, "Geology of Unalaska Island and Adjacent Insular Shelf, Aleutian Islands, Alaska," Bulletin No. 1028-S, US Geological Survey.

Forster, H. L., 1970, "Reconnaissance Geologic Map of Tanacross Quadrangle, Alaska," Miscellaneous Geologic Inventory Map No. I-593, US Geological Survey.

Gates, O., Powers, H. A., and Wilcox, R. E., 1971, "Geology of the Near Islands," Bulletin No. 1028-U, US Geological Survey.

Geochemical Exploration, 1972, Special Vol. No. 11, Canadian Institute of Mining and Metallurgy, pp. 67-77.

Godwin, C.I., 1975, "The Casino Porphyry Copper Deposit," Ph.D. Thesis, University of British Columbia, Vancouver, B.C., Canada, 238 pp.

Grantz, A., and Kirschner, E. E., 1975, "Tectonic Framework of Petroliferous Rocks in Alaska," Open File Report, US Geological Survey.

Grow, J. A., 1973, "Crustal and Upper Mantle Structure of the Central Aleutian Arc," Bulletin, Geological Society of America, Vol. 84, pp. 2169-2192.

Hawley, C.C., et al., 1969, "Results of Geological and Geochemical Investigations in an Area Northwest of the Chulitna River, Central Alaska Range," Circular No. 617, US Geological Survey, p. 19.

Herreid, G., and Kaufman, M. A., "Geology of the Dry Pass Area, Southeastern Alaska," Geology Report No. 7, Alaska Div. of Mines and Minerals, p. 12.

Hollister, V. F., Anzalone, S. A., and Richter, D. H., 1975, "Porphyry Copper Belts of Southern Alaska and Contiguous Yukon Territory," Technical Paper No. 22, Canadian Institute of Mining and Metallurgy Annual Meeting.

Kaufman, M. A., 1964, "Geology and Mineral Deposits of the Denali-Maclaren River Area, Alaska," Geology Report No. 4, Alaska Div. of Mines and Minerals, p. 15.

Lanphere, M. A., and Reed, B. L., 1973, "Timing of Mesozoic and Cenozoic Plutonic Events in Circum Pacific North America," Bulletin, Geological Society of America, Vol. 84, pp. 3773-3782.

Larson, R. L., and Chase, C. G., 1972, "Late Mesozoic Evolution of the Western Pacific Ocean," Bulletin, Geological Society of America, Vol. 83, p. 3645.

Lathram, E. H., and Gryc, G., 1972, "Metallogenic Significance of Alaskan Geostructures Seen from Space," Proceedings, 8th International Symposium on Remote Sensing, University of Michigan Press, Ann Arbor, Mich.

Lowell, J. D., and Guilbert, J. M., 1970, "Lateral and Vertical Alteration Mineralization Zoning in Porphyry Ore Deposits," Economic Geology, Vol. 65, pp. 373-408.

MacKevett, E. M., Jr., et al., 1971, "Mineral Resources of Glacier Bay National Monument, Alaska," Professional Paper No. 632, US Geological Survey, p. 90.

Marlow, M. S., et al., 1973, "Tectonic History of the Central Aleutian Arc," Bulletin, Geological Society of America, Vol. 84, p. 1555.

Miller, T. P., Patton, W. W., and Lanphere, M. A., 1966, "Preliminary Report on a Plutonic Belt in West-Central Alaska," Professional Paper No. 550-D, US Geological Survey, pp. D158-D162.

Miller, T. P., and Ferrians, O. J., 1968, "Suggested Areas for Prospecting in the Central Koyukuk River Region, Alaska," Circular No. 570, US Geological Survey.

Miller, T. P., and Elliott, R. L., 1969, "Metalliferous Deposits Near Granite Mountain, Eastern Seward Peninsula, Alaska," Circular No. 614, US Geological Survey.

Miller, T. P., 1970, "Petrology of the Plutonic Rocks of West-Central Alaska," Open File Report, US Geological Survey, p. 132.

Miller, T. P., 1972, "Potassium-Rich Alkaline Intrusive Rocks of Western Alaska," Bulletin, Geological Society of America, Vol. 83, pp. 211-2128.

Moffit, F. H., and Knoff, A., 1910, "Mineral Resources of the Nabesna-White River District, Alaska," Bulletin No. 417, US Geological Survey, p. 64.

Moore, C., 1973, "Complex Deformation of Cretaceous Trench Deposits, Southwestern Alaska," *Bulletin,* Geological Society of America, Vol. 84, pp. 2005-2020.

Naugler, F. P., and Wageman, J. M., 1973, "Gulf of Alaska: Magnetic Anomalies, Fracture Zones and Plate Interaction," *Bulletin,* Geological Society of America, Vol. 84, pp. 1575-1584.

Patton, W. W., 1970, "Preliminary Geologic Investigations in the Kanuti River Region, Alaska," Bulletin No. 1312-J, US Geological Survey.

Patton, W. W., and Csejtey, B., 1971, "Preliminary Geologic Investigations of Western St. Lawrence Island, Alaska," Professional Paper No. 684-C, US Geological Survey.

Patton, W. W., 1973, "Reconnaissance Geology of the Northern Yukon-Koyukuk Province, Alaska," Professional Paper No. 774-A, US Geological Survey, pp. 1-17.

Pitman, W. C., and Hayes, D. E., 1968, "Seafloor Spreading in the Gulf of Alaska," *Journal of Geophysical Research,* Vol. 73, p. 6571.

Reed, B. L., and Elliott, R. L., 1968, "Geochemical Anomalies and Metalliferous Deposits between Windy River and Post River, Southern Alaska Range," Circular No. 569, US Geological Survey.

Reed, B. L., and Lanphere, M. A., 1969, "Age and Chemistry of Mesozoic and Tertiary Plutonic Rock in South-Central Alaska," *Bulletin,* Geological Society of America, Vol. 80, pp. 23-44.

Reed, B. L., and Elliott, R. L., 1970, "Reconnaissance Geologic Map, Analyses of Bedrock and Stream Sediment Samples and an Aeromagnetic Map of Parts of the Southern Alaska Range," Open File Report, US Geological Survey.

Reed, B. L., and Lanphere, M. A., 1972, "Generalized Geologic Map of the Alaska-Aleutian Range Batholith Showing K-Ar Ages of the Plutonic Rocks," Miscellaneous Field Map No. MF-372, US Geological Survey.

Reed, B. L., and Lanphere, M. A., 1973, "Alaska-Aleutian Range Batholith: Geochronology, Chemistry and Relation to Circum-Pacific Plutonism," *Bulletin,* Geological Society of America, Vol. 84, pp. 2583-2610.

Richter, D. H., 1966, "Geology of the Slana District, Southcentral Alaska," Geology Report No. 21, Alaska Div. of Mines and Minerals, p. 51.

Richter, D. H., and Jones, D. L., 1973, "Reconnaissance Geologic Map of the Nabesna A-2 Quadrangle, Alaska," Miscellaneous Geologic Inventory Map No. I-749, US Geological Survey.

Richter, D. H., 1974, "Reconnaissance Geologic Map of the Nabesna A-4 Quadrangle, Alaska," Miscellaneous Geologic Inventory Map No. I-830, US Geological Survey.

Richter, D. H., Matson, N. A., Jr., and Schmoll, H. R., 1974, "Reconnaissance Geologic Map of the Nabesna A-1 Quadrangle, Alaska," Miscellaneous Geologic Inventory Map No. MF 689, US Geological Survey (in press).

Rose, A. W., 1967, "Geology of the Upper Chistochina River Area, Mt. Hayes Quadrangle, Alaska," Geology Report No. 28, Alaska Div. of Mines and Minerals, p. 39.

Sainsbury, C. L., 1969, "Geology and Ore Deposits of the Central York Mountains, Western Seward Peninsula, Alaska," Bulletin No. 1287, US Geological Survey, p. 101.

Smith, T. E., 1971, "Bedrock Geology of the Mount Hayes A-6 Quadrangle," Annual Report, US Geological Survey, Alaska Div., pp. 28-31.

Stacey, R. A., "Gravity Anomalies, Crustal Structure and Plate Tectonics in the Canadian Cordillera," *Canadian Journal of Earth Sciences,* Vol. 10, p. 615.

Templeman-Kluit, D. J., 1973, "Reconnaissance Geology of Aeshihik Lake, Snag and Part of the Stewart River Map-Areas, West Central Yukon," Open File Report No. 161, Geological Survey of Canada.

Templeman-Kluit, D. J., 1976, "The Yukon Terrane," *Bulletin,* Geological Society of America, Vol. 87, pp. 1343-1357.

Warner, et al., 1961, "Iron and Copper Deposits of Kasaan Peninsula, Prince of Wales Island, Southeastern Alaska," Bulletin No. 1090, US Geological Survey, p. 136.

4

Porphyry Copper Deposits of the Northern Cordilleran Orogen

CONTENTS

INTRODUCTION

This chapter summarizes characteristics of porphyry copper deposits within the Cordilleran orogen east of the Coast Range plutonic complex of the Yukon and British Columbia and south to the Columbia River plateau and the Idaho batholith. Porphyry copper deposits in this area reflect a geologic environment different from that found for mineralized intrusions elsewhere in the Cordilleran orogen, both to the south in the United States and Mexico and to the west in Alaska. A section on the Tertiary Cascade porphyry copper province is included herein. The Cascade province is considered separately because its deposits display a unique mineralogy and age of mineralization. That portion of the Cordilleran orogen described in this chapter is shown in Fig. 32, with the Cascades shown in Fig. 39 (see p. 114).

The regional geology of this area has been described by Monger, et al. (1972), Hollister (1974), Wolfhard and Ney (1974),

Field, et al. (1974), and Stacey (1974). Each offer slightly differing views on the post-Mississippian evolution of the geology in this porphyry copper province. The descriptive geology in this chapter has benefited from many discussions with J. E. Armstrong and J. W. H. Monger.

Two separate petrologic models have been used popularly to categorize porphyry copper deposits in this area, a diorite and a quartz monzonite model. The diorite model, which involves a mineralized quartz-free pluton, has also been called the alkalic type (*CIM Special Volume* 15, 1976, and Soregaroli, 1975) and the syenite suite (Sutherland Brown, et al., 1971). The quartz monzonite model has also been named the calc-alkalic type (*CIM Special Volume* 15, 1976) and includes mineralized quartz-bearing plutons compositionally ranging from quartz diorite to granite, with the igneous phase closest to ore in time and space commonly being quartz monzonite.

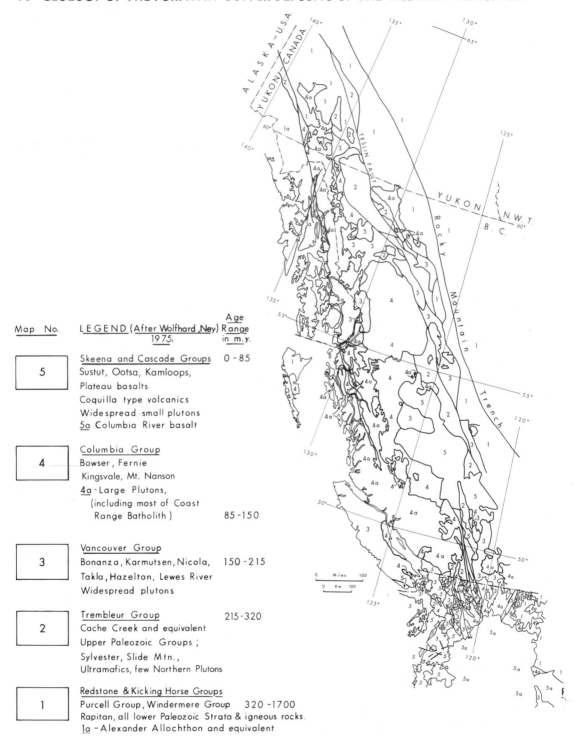

Map No.	LEGEND (After Wolfhard ,Ney) 1975.	Age Range in m.y.
5	Skeena and Cascade Groups Sustut, Ootsa, Kamloops, Plateau basalts Coquilla type volcanics Widespread small plutons 5a Columbia River basalt	0 - 85
4	Columbia Group Bowser, Fernie Kingsvale, Mt. Nanson 4a - Large Plutons, (including most of Coast Range Batholith)	85 -150
3	Vancouver Group Bonanza, Karmutsen, Nicola, Takla, Hazelton, Lewes River Widespread plutons	150 -215
2	Trembleur Group Cache Creek and equivalent Upper Paleozoic Groups; Sylvester, Slide Mtn., Ultramafics, few Northern Plutons	215-320
1	Redstone & Kicking Horse Groups Purcell Group, Windermere Group Rapitan, all lower Paleozoic Strata & igneous rocks. 1a - Alexander Allochthon and equivalent	320 -1700

Fig. 32. Geologic index map, northern Cordilleran orogen. Geologic map showing metallogenic epochs of Wolfhard and Ney (1975). These metallogenic epochs coincide with stratigraphic-tectonic units and are used in that sense in the text. Porphyry copper deposits have been dated from the youngest three map units used in this summary (*modified from Wolfhard and Ney, 1974*).

The quartz monzonite model has been further divided into granitic pluton type (Field, et al., 1974) and stock type (including the volcanic and phallic types of Sutherland Brown, 1972). The stock type contains porphyry deposits associated with small plutons or dikes. Some consideration is given in this chapter to modifying the Lowell and Guilbert (1970) model so that quartz monzonite model deposits of this area can be described by it.

Table 11 summarizes data on the largest known diorite-type deposits within the area. Table 12 is a compilation of data on the most important known granitic pluton-type deposits, and Table 13 lists the most important stock types. Deposits listed in these tables are shown in Fig. 33. Molybdenum porphyries (Clark, 1972) are omitted from consideration in this chapter as are large copper-bearing skarn deposits.

GEOLOGIC SETTING

Late Paleozoic and Mesozoic geologic development of the Canadian Cordillera (Fig. 32) is unusual for the eastern Pacific rim. Because geologic evolution from Mississippian time on was particularly influential in the development of deposits in this porphyry copper province, the post-Mississippian geology of this portion of the Cordilleran orogen is discussed in some detail. The geology in this chapter largely follows Wolfhard and Ney (1974) and *CIM Special Volume* 15 (1976) and is basically the same as that found in White (1959) and Monger (1975).

Wolfhard and Ney (1974) propose and name a number of metallogenic epochs. These epochs also have been used as stratigraphic-tectonic elements and are used in that fashion in this chapter. The map units in Fig. 32 correspond to these metallogenic epochs.

Map Unit One: Map unit one includes all the cratonic, platform, and shelf deposits and any associated igneous rock that predate 360 m.y. in the area of this map unit. These stratigraphic-tectonic elements are generalized and equated to the Redstone and Kicking Horse metallogenic epochs by Wolfhard and Ney (1974). They comprise the cratonic sialic crust that was rifted, dextrally faulted, folded, and intruded in ensuing time. Porphyry copper deposits formed as one consequence of successive younger tectonic events. The Alexander terrane of Berg, et al. (1972) is set aside as a unit in Fig. 32, as are the Lower Paleozoic and Precambrian rocks of the Cascades.

Map Unit Two: Map unit two is the Trembleur metallogenic epoch of Wolfhard and Ney (1974) and consists of the stratigraphic-tectonic element formed in the 360 to 215 m.y. period. It includes Cache Creek and equivalent groups (e.g., Sylvester and Slide Mountain) that occur in the area of this map unit. As described by Wheeler, et al. (1972), Cache Creek includes all the elements of oceanic crust. The tectonic significance of oceanic crust between the Alexander terrane and the Cascade Paleozoics on the west and the North American craton on the east is not entirely clear, but seems to imply western tectonic transport for rocks represented by map unit 1a, Alexander terrane, which has been identified west of the oceanic crust. Carbonatites near the western margin of the cratonic block of Mississippian(?) age may be speculated to be coeval with earliest development of oceanic crust and to coincide with early distensional tectonics.

Map Unit Three: Map unit three (Fig. 32) is the tectonic-stratigraphic element that includes dominantly marine volcanic assemblages with dates from 215 to 150 m.y. It is called the Vancouver metallogenic epoch by Wolfhard and Ney (1974) and includes Nicola, Takla, and Hazelton groups on the mainland and Bonanza and Karmutsen formations on Vancouver Island. The Triassic and pre-Upper Jurassic Bonanza and Karmutsen calc-alkalic, marine volcanic, and sedimentary rocks on Vancouver Island are similar in some respects to the stratigraphically equivalent Nicola-Hazelton-Takla groups. These latter, however, include alkalic members that developed in a distensional environment whereas volcanic formations on Vancouver Island do not have alkalic flows and

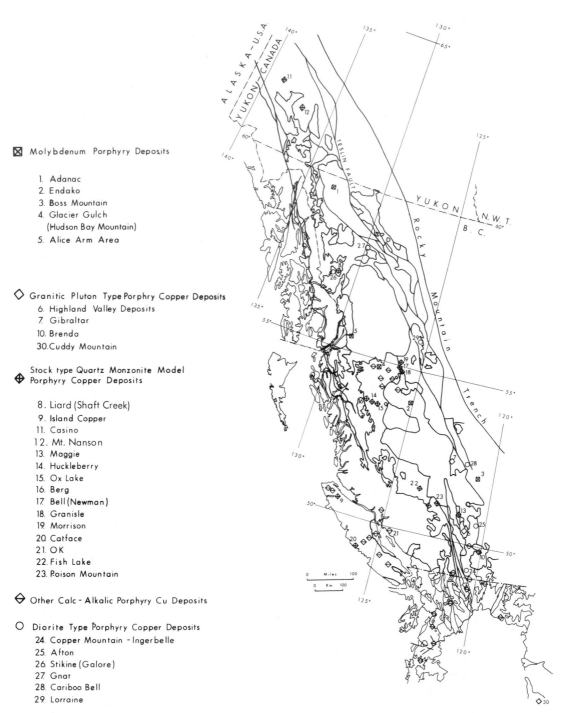

⊠ Molybdenum Porphyry Deposits

   1. Adanac
   2. Endako
   3. Boss Mountain
   4. Glacier Gulch
     (Hudson Bay Mountain)
   5. Alice Arm Area

◇ Granitic Pluton Type Porphry Copper Deposits
   6. Highland Valley Deposits
   7. Gibraltar
  10. Brenda
  30. Cuddy Mountain

⊕ Stock type Quartz Monzonite Model
Porphyry Copper Deposits

   8. Liard (Shaft Creek)
   9. Island Copper
  11. Casino
  12. Mt. Nanson
  13. Maggie
  14. Huckleberry
  15. Ox Lake
  16. Berg
  17. Bell (Newman)
  18. Granisle
  19. Morrison
  20. Catface
  21. OK
  22. Fish Lake
  23. Poison Mountain

◆ Other Calc-Alkalic Porphyry Cu Deposits

○ Diorite Type Porphyry Copper Deposits
  24. Copper Mountain - Ingerbelle
  25. Afton
  26. Stikine (Galore)
  27. Gnat
  28. Cariboo Bell
  29. Lorraine

Fig. 33. Location of major deposits. Major known porphyry deposits are located on the geologic outline used in Fig. 31. Porphyry molybdenum deposits are included for those who wish to refer to them (*compiled from various British Columbia Dept. of Mines sources*).

Table 11. Diorite Model Porphyry Copper Deposits of the Canadian Cordillera

Name	Location	Intrusion Close to Ore	Intruded	Age in m.y.	Alteration Sequence from Center	% Py in Pot Zone	Size of Py Halo	Structure
Afton	50°39'; 120°30'	Dio Sye	Trias Vol	198	Prop	1	Erratic	Stockwork
Cariboo Bell	52°30'; 121°38'	Sye Mon	Trias Vol	—	Pot-Phy-Prop	2	1.0 x 1.1 km	Stockwork
Copper Mountain	49°21'; 121°34'	Sye	Trias Vol	193	Pot-Prop	3	2.0 x 4.0 km	Stockwork
Duckling Creek (Lorraine)	55°55'; 125°26'	Sye	Trias Vol	170	Pot-Prop	1	2.0 x 3.0 km	Stockwork
Galore Creek (Stikine)	57°07'; 131°26'	Sye	Trias Vol	182	Pot-Phy-Prop	3	4.0 x 5.0 km	Breccia
Rayfield River	51°15'; 121°04'	Mon Dio	Trias Vol	—	Pot-Prop	1	2.0 x 2.6 km	Stockwork
Gnat Lake	58°11'; 129°51'	Dio	Trias Vol	—	Pot-Prop	2	1.2 x 1.5 km	Stockwork

Qtz: Quartz
Mon: Monzonite
Dio: Diorite
Grdr: Granodiorite

Por: Porphyry
Trias: Triassic
Vol: Volcanics
Sed: Sediments

Km: Kilometers
Sye: Syenite

Pot: Potassic
Phy: Phyllic
Arg: Argillic
Prop: Propylitic

Table 12. Granitic Pluton Type Porphyry Copper Deposits of the Northern Cordilleran Orogen

Name	Location	Intrusion Close to Ore	Intruded Rock	Age m.y.	Zoning Sequence from Center	% Py in Phy Zone	Size of Py Zone	Structure
Guichon Batholith (Bethlehem, Lornex, Valley)	50°29'; 121°00'	Qtz Mon Por	Trias Vol	200	Pot-Phy-Arg-Prop	3-5	Erratic	Stockwork
Brenda	49°52'; 120°01'	Grdr	Trias Vol	140(?)	Pot-Prop	1	Erratic	Stockwork
Gibraltar	52°31'; 122°16'	Qtz Diorite	Trias Vol	204	Pot-Phy-Prop	1	Erratic	Stockwork
Cuddy Mtn.	45°10'; 116°15'	Grdr Por	Trias Vol	200	Pot-Phy-Prop	2	Erratic	Bx and Stwk

Qtz: Quartz
Mon: Monzonite
Dio: Diorite
Grdr: Granodiorite

Por: Porphyry
Trias: Triassic
Vol: Volcanics
Sed: Sediments

Km: Kilometers
Bx: Breccia
Stwk: Stockwork

Pot: Potassic
Phy: Phyllic
Arg: Argillic
Prop: Propylitic

Table 13. Stock Type Porphyry Copper Deposits of the Northern Cordilleran Orogen

Name	Location	Intrusion Close to Ore	Intruded Rock	Age m.y.	Zoning Sequence from Center	% Py in Phy Zone	Size of Py Zone	Structure
Island Copper	59°36'; 127°28'	Qtz Por	Trias Vol	153	Pot-Phy-Arg-Prop	4	1.6 x 2.2 km	Stockwork
Liard (Shaft Creek)	57°19'; 130°50'	Qtz Dio Por	Trias Vol	182	Pot-Phy-Prop	3	1.0 x 2 km	Stockwork
Catface	49°15'; 125°58'	Qtz Mon Por	Pal(?) Sed	48	Pot-Phy-Arg-Prop	3	1.1 x 1.4 km	Stockwork
Granisle	54°58'; 126°19'	Qtz Mon Por	Trias Vol	51	Pot-Phy-Arg-Prop	4	1.3 x 1.3 km	Breccia
Huckleberry	53°49'; 127°10'	Qtz Mon Por	Trias Vol	80	Pot-Phy-Arg-Prop	5	1.3 x 1.3 km	Stockwork
Ox Lake	53°40'; 127°03'	Qtz Mon Por	Trias Vol	83	Pot-Phy-Prop	4	1.3 x 1.3 km	Stockwork
Bell (Newman)	55°00'; 126°14'	Qtz Dio Por	Trias Vol	52	Pot-Phy-Arg-Prop	4	1.5 x 2.1 km	Stockwork
Berg	53°47'; 127°28'	Qtz Mon Por	Trias Vol	50	Pot-Phy-Arg-Prop	4	1.5 x 1.6 km	Stockwork
Bond Creek	62°15'; 142°46'	Qtz Dio Por	Trias Vol	109	Pot-Phy-Arg-Prop	5	4.0 x 6.1 km	Breccia
Casino	62°45'; 138°48'	Qtz Por	Pal(?) Sed	70	Pot-Phy-Arg-Prop	6	3.0 x 4.1 km	Breccia
Poison Mtn.	49°21'; 120°34'	Qtz Dio Por	Cret Sed	?	Pot-Phy-Prop	1	1.0 x 1.7 km	Stockwork
Maggie	50°40'; 121°20'	Qtz Mon Por	Trias Vol	61	Pot-Phy-Arg-Prop	4	1.5 x 3.7 km	Stockwork
Fish Lake	50°46'; 133°04'	Qtz Dio Por	Trias Vol	77	Pot-Phy-Prop	2	2.5 x 3.6 km	Stockwork
Morrison	55°02'; 127°08'	Grdr Por	Trias Vol	52	Pot-Phy-Prop	3	2.3 x 2.8 km	Stockwork

Pot: Potassic
Phy: Phyllic
Arg: Argillic
Prop: Propylitic

Qtz: Quartz
Mon: Monzonite
Dio: Diorite
Grdr: Granodiorite

Por: Porphyry
Trias: Triassic
Vol: Volcanics
Sed: Sediments

Pal: Paleozoic
Perm: Permian
Km: Kilometers
Cret: Cretaceous

have not been assigned such a setting (Stacey, 1974).

Nicola, Takla, and Hazelton rocks of map unit three occur east of the Alexander terrane (or allochthon), both west and east of the Cache Creek ophiolite (map unit two), and west of the North American craton. Their base is only rarely exposed, but Preto (1975) reports Triassic and Lower Jurassic fossils occur in Nicola volcanogenic sediments associated with marine tholeiites, feldspathoid-bearing alkalic and subalkalic tholeiites, and volcanic differentiation products (e.g., dacite). These volcanic rocks include coeval calc-alkalic, alkalic, and subalkalic units.

The Triassic and Lower Jurassic volcanic rocks, regardless of setting, are the best host for most porphyry copper deposits in the Canadian Cordillera irrespective of age and are thus an important factor in prospecting for and positioning of these deposits.

Intrusions through and comagmatic with the Triassic Nicola-Takla-Hazelton or equivalent volcanic assemblages east of the Alexander allochthon do not appear to be typical products of an arc-trench environment. Sierra Nevada type batholiths are missing, as is Franciscan type trench melange. Groups of alkalic and dioritic plutons occur comagmatic with undersaturated volcanics in remarkably linear zones over many miles (Fox, 1975). Normal faults active over a considerable period of time have not only controlled the position of the dioritic and alkalic intrusive centers but the distribution of associated alkalic volcanic rocks as well.

On the other hand, Triassic and Lower Jurassic calc-alkalic plutons comagmatic with extrusives of the Nicola and Takla groups are typically zoned with a basic outer shell giving way gradationally inwards to a more alkalic and silicic core. These zoned plutons extend from Texas Creek in Alaska to Cuddy Mountain in Idaho (Field, et al., 1974). The large zoned calc-alkalic plutons that have copper associated with their core phases are the granitic pluton type porphyry copper of Field, et al. (1974). The diorite type porphyry copper is the alkalic, commonly zoned,

diorite-monzonite-syenite pluton where copper occurs in or near the younger differentiates. Therefore, porphyry copper deposits may have a quartz-bearing quartz monzonite (calc-alkalic) or quartz-free diorite (including alkalic) host.

Why two distinct suites of plutons (alkalic and calc-alkalic) occur in the same general area, each originating simultaneously with the same set of volcanics but presumably from a different crustal depth although in response to the same tectonic regime, is not clear. The alkalic volcanic rocks and plutons form an unusual setting for porphyry copper deposits in that they seem to satisfy the criteria for a spreading center, and the zoned mineralized plutons comagmatic with them are probably the roots or magma chambers of old alkalic volcanic centers.

Unmineralized batholithic calc-alkalic plutons with dates that range from 185 to 165 m.y. have been found in the Yukon east of the Shakwak fault. Middle Jurassic batholithic plutons have also been dated in the southern Cordillera near the 49th parallel.

The Columbian orogeny (Wheeler, et al., 1972) apparently closed this period of mixed alkalic and calc-alkalic volcanism and plutonism.

Map Unit Four: Map unit four (Fig. 32) includes various rocks that comprise the stratigraphic-tectonic unit formed in the interval 150 to 80 m.y. It is called the Columbia metallogenic epoch by Wolfhard and Ney (1974) and included with this stratigraphic tectonic element are the Bowser, Fernie, Kingsvale, and Mount Nanson groups as well as most of the Coast Range plutonic complex. The Granvina-Nutzotin arc-trench accumulations in Alaska as well as back arc type basin accumulations of this age east of the batholiths infer that the plutonic complex is the axis of an island-arc volcanic belt.

Larson and Pitman (1972) have found a long period of normal magnetic polarity to extend from 110 to 80 m.y. Wanless (1969) reports numerous radiometric ages for batholiths in the Canadian Cordillera from this period. Wanless (1969) also reports common dates from the coastal batholith in the 50-42 m.y. interval, although in terms of volume of batholithic material formed the 110-80 m.y. interval is the most important.

Larson and Pitman (1972) interpret the magnetic quiet intervals as periods of rapid sea-floor spreading. Subduction during these intervals was therefore more rapid and batholithic plutonism increased. Concurrent with the increase in subduction rates, compressional tectonics were dominant. Major folding and thrusting on the continents but not major transcurrent faulting have been stratigraphically dated to coincide with the magnetic quiet zones of the oceanic basins (Wheeler, et al., 1972).

As noted on Tables 11, 12, and 13, very few dated porphyry copper deposits have ages occurring within this period of batholithic activity. Because porphyry copper deposits uniformly exhibit extensional structural conditions, the compressive tectonic environment accompanying more rapid subduction and corresponding to periods of more pronounced batholithic activity should not be favorable to their formation. Dating on deposits in the Canadian Cordillera confirms this negative correlation.

Map Unit Five: Map unit five, the Skeena and Cascade metallogenic epochs of Wolfhard and Ney (1974), represents the stratigraphic-tectonic element formed from 80 m.y. to the present. A significant break at about 42 m.y. separates the older stock and small batholith-dominant igneous activity from the plateau bimodal volcanism more common in the younger Cascade metallogenic epoch. These epochs include the Sustut, Ootsa, and Kamloops groups and Coquihalla type volcanics. A large number of K-Ar dates within the range 50-42 m.y. are now available for batholithic plutons in the Coast Range plutonic complex. These have made some other K-Ar dates younger and confused somewhat the distribution of plutons by age.

Most porphyry copper deposits dated in this epoch are diapric. They occur in areas where regional stress was characterized and dominated by strike-slip faulting (Monger, et al., 1972). The host for the Tertiary porphyry copper deposits appears to have been largely Triassic volcanic rocks (Fig. 35). Tertiary porphyry copper deposits mostly penetrated the Lower Mesozoic volcanic terrane, which included the older comagmatic diorite and granitic pluton type porphyry deposits. The reappearance in the Tertiary of porphyry copper deposits in the Triassic volcanic terrane is unusual in the eastern Pacific rim.

Porphyry copper deposits do not commonly occur outside the Lower Mesozoic volcanic rocks whereas porphyry molybdenum type occurrences are known within the Coast Range plutonic complex and with some larger plutons (Sutherland Brown, et al., 1971). The absence of porphyry copper deposits combined with the presence of porphyry molybdenum deposits in the Coast Range plutonic complex could be ascribed to a difference in type of crust. This speculation is not yet supported by adequate geochemical or geophysical data.

Structure

Fig. 34 compiles strike-slip faults mapped by the Geological Survey of Canada, the US Geological Survey, the British Columbia Dept. of Mines, and the Washington State Dept. of Natural Resources within the area of this chapter. It is modified after Seraphim and Hollister (1975). Thrust faults are generalized or omitted in the vicinity of the 49th parallel, the Alaska-Canada border, and the eastern part of this section (e.g., Rocky Mountain thrust belt) because porphyry deposits are only rarely associated with this type of fracture.

The fracture pattern shown in Fig. 34 is compatible with that which would develop if a Pacific plate moved north with respect to the North American plate. The major shears trend about N35W (e.g., Pinchi, Teslin, and Tetlin faults) and the major tensional fractures trend nearly north-south. Together these form the Pacific plate-North American plate stress diagram (Fig. 34). The structural elements conjugate to these two fracture directions could also include shears trending about N35E. The northeast set is locally identifiable

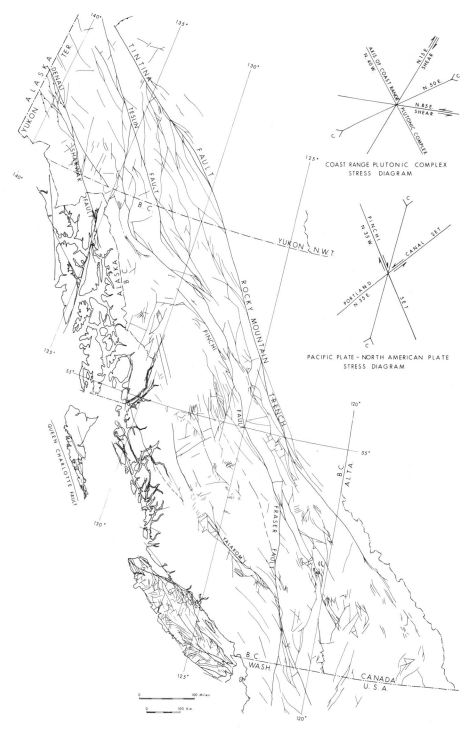

Fig. 34. Structure in the northern Cordilleran orogen. Major strike-slip faults known to exist in the northern Cordilleran orogen are shown on this map. Most porphyry deposits shown in Fig. 32 fall near a major strike-slip fault. Thrust faults, fold axes, and plutonic rocks have been omitted to simplify the map (*compiled from various British Columbia Dept. of Mines and Geological Survey of Canada sources and from Seraphim and Hollister, 1975*).

although not as well developed as the Pinchi. The north-south tensional fractures formed (as in the case of the Highland Valley) simultaneously with the N35W compressive shears having right lateral movement and therefore are related to the causative force of the strike-slip tectonics.

The northwest shears form a branching inter-fingered network of transcurrent faults whose west blocks invariably have moved north. Fault displacements are generally younger on the westernmost shears. The amount of movement on these displacements has not been determined either separately or as a whole. On the other hand, movement of the Pacific plate relative to the North American craton could be hundreds of kilometers. Aggregate movement on all northwest shears between Vancouver Island and the North American craton also may have been hundreds of kilometers. It is clear from this reasoning that anything west of the Rocky Mountain trench may be allochthonous with respect to rocks east of the trench and the numerous branching northwest—trending right lateral faults in reality may form one large fault zone separating the craton from the Pacific plate. Most porphyry deposits occur close to one of the major strike-slip faults and the spatial association of the two suggest a genetic connection. It would be logical to assume that these faults provided access to the upper crust for the porphyry magma.

Thrust faults in this area generally are compatible with a separate and distinct N50E compressive force. The tension fractures and strike-slip faults associated with N50E foreshortening make up the Coast Range plutonic complex stress diagram (Seraphim and Hollister, 1975) in Fig. 34. Because the Coast Range plutonic complex and the fold axes in its vicinity trend N40W, it would seem that the compressive force that resulted in the thrusts was genetically related to Coast Range plutonism and folding. Subduction during parts of the Jurassic and Cretaceous appears responsible for the northeast-trending compressive force (Monger, et al., 1972) as well

as the N40W fold axes and intrusion-related manifestations of northeast compression. Vertical shears related to this compressive force that trend approximately N85E (with the north block moving west) and N15E are identifiable within the Cordilleran orogen, but these lack the continuity, number, or significance of the N35W set of the Pacific plate-North American plate stress diagram.

Porphyry copper deposits appear timed to occur most frequently during periods of strike-slip tectonics rather than during periods of northeast compression (Hollister, 1974). Since few deposits actually occur directly on strike-slip faults, however, the faults characterize the tectonic style within which the porphyry copper deposit is most likely to develop.

The N50E-oriented compressive force would appear to be episodic and dates on batholithic plutons within the Coast Range plutonic complex may be related to peak periods of compression. If this conjecture is accepted then the compressive force oriented N50E may have been most active in the periods 165 to 160, 110 to 80, and 50 to 42 m.y.

The compressive force oriented north-south may also have been episodic. The oceanic crust of map unit two (Cache Creek) could be considered a product of both rifting and strike-slip faulting related to northerly movement of the Pacific plate relative to the North American. Should this be more conclusively demonstrated eventually, the north-south compression may have operated as far back as the interval from Mississippian to mid-Triassic. If dating on the porphyry copper deposits developed on strike-slip faults may be used to date fault displacements, it is possible that porphyry dates could be used to infer timing of the north-south force. If this speculation is accepted, the north-south force could have been active in the periods 215 to 170, 153 to 140, and 83 to 65 m.y. Based on stratigraphic evidence, it appears to have been effective through most of Early Tertiary.

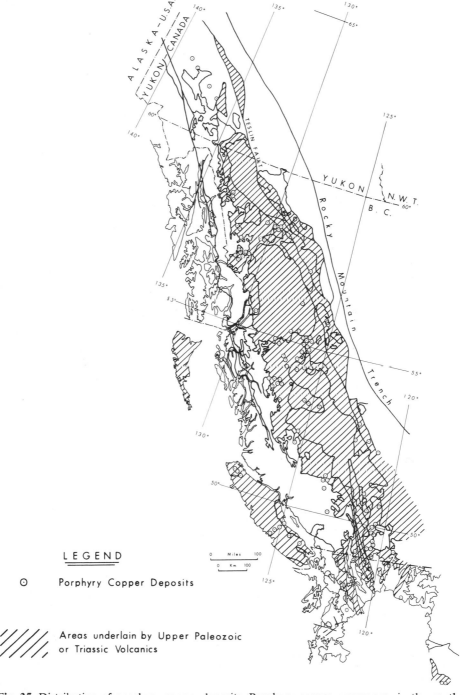

Fig. 35. Distribution of porphyry copper deposits. Porphyry copper occurrences in the northern Cordilleran orogen regardless of economic potential are shown on this map together with those areas of the subsurface that include with their sialic crust Permian or Triassic basic marine volcanics. Relatively few occurrences are known outside the areas where basic marine volcanics are suspected in the crust (*modified from Hollister, 1974*).

Discussion

The appearance of Triassic-Lower Jurassic diorite and granitic pluton type and Tertiary quartz monzonite stock type porphyry copper deposits in the same terrane justifies a more detailed examination of crustal setting.

A step-by-step growth in recognition of the Alaskan Paleozoic or Alexander allochthon was enhanced with the discovery of an anomaly in Permian fossils by Ross (1967). This anomaly was further developed by Monger and Ross (1971) and Jones, et al. (1972). Their investigations have led to recognition that pre-Mississippian rocks west of the outcropping Permian-Triassic-Lower Jurassic sedimentary and volcanic rocks— i.e., west of the stratigraphically complex and extensive Cache Creek-Hazelton-Takla-Nicola groups—are allochthonous with respect to the North American craton.

Tectonic transport is the logical vehicle for formation of the tectonic element represented by map unit two oceanic crust (Pennsylvanian, Permian, and Lower Triassic—Wheeler, et al., 1972). Its development between the Alexander terrane and the North American craton offers one solution to the problem of why the Alexander terrane (Berg, et al., 1972) or allochthon (Jones, et al., 1972) occurs. The allochthon includes rocks at least as young as Upper Triassic and appears to have moved, possibly intermittently, during the period from the Mississippian until at least the Upper Triassic. Berg, et al. (1972) imply that movement should have ceased by the Upper Jurassic because strata of this age (their Gravina-Nutzotin belt) overlie allochthonous Paleozoic crust that does not itself appear tectonically transported. Tectonic transport may actually have ceased by the time of the Middle Jurassic Columbian orogeny (Wheeler, et al., 1972). The absence of large-scale batholithic plutonism from the Mississippian to the Middle Jurassic therefore is significant.

Paleozoic rocks attached to the western margin of the craton (map unit one), where they are in contact with the Triassic and Lower Jurassic (Nicola-Takla-Hazelton of map unit three, Fig. 32) display an outcrop pattern typically found in rift margins. Fox (1975) cites additional evidence of rifting during the Takla alkalic volcanic episode. The contacts suggest a high probability that initial separation of the allochthon was by rifting, although much later displacement clearly appears to have been accomplished at least partially by right lateral movement on several of the strike-slip faults known to have developed prior to the Tertiary. The association of alkalic plutonism, volcanism, and large-scale normal faulting implies that a period of Lower Mesozoic distentional tectonism largely in a marine environment developed during this period of strike-slip tectonics. Preto (1975) also cites evidence that Triassic rift structures that were developing concomittantly with Nicola volcanism (map unit three of Fig. 32) were compatible with strike-slip tectonics for adjacent terranes.

K-Ar dating of the long, but less than 8 km (5 mile) wide band of blue schist metamorphic facies rocks within the Pinchi fault have indicated ages from 216 to 211 m.y. (Paterson and Harakal, 1974). The regional geologic setting clearly suggests that strike-slip movement took place on this high angle fault during metamorphism, although Paterson and Harakal (1974) mention the possibility of thrusting. The high-pressure low-temperature mineralogy of the rocks could have formed under either strike-slip or thrust conditions but the case for dextral fault displacement is easier to understand in light of evidence cited by Preto (1975) and Fox (1975).

The Paleozoic-Precambrian rocks of Washington's Cascades (included as group 1a in Fig. 32) are bounded on the east by formations lithologically identical to and of the same age as the Cache Creek-Hazelton-Takla-Nicola groups that separate the Alaskan Paleozoic allochthon from the North American craton. Should presently undated metamorphic rocks along the west margin of British Columbia's Coast Range plutonic complex and between the Alexander allochthon and the Cascade Paleozoic eventually be shown to include Paleozoic rocks, then the Paleozoics from the Cascades to Alaska may be demonstrated as one large allochthon.

The allochthonous crust differs from strata of the same age that forms part of the nearby North American craton. The allochthon has been displaced northwesterly relative to the craton. It also represents the continental margin accumulations, whereas the craton contains more typical shelf and platform strata.

Porphyry copper deposits that formed in the interval 205-170 m.y. therefore developed at the end of a long period of plate distension. Included with these deposits are all of the diorite and most of the granitic pluton type. Within this period of Lower Mesozoic distensional tectonics, only diorite and granitic pluton type deposits are known to have formed. These developed comagmatic with their extrusive hosts as volcanic centers. The conditions of rifting described by Fox (1975) and Preto (1975) for the porphyry environment were preceded by a long history of rifting and strike-slip faulting.

Studies in Alaska that are cited in this volume indicate that marine volcanic accumulations now located along the continental margin and south of the Denali fault of Pennsylvanian to pre-Upper Jurassic age also may be in part allochthonous with respect to the North American craton. These orogenic accumulations added to the continental margin (the Taku-Skolai terrane of Berg, et al., 1972) are identical to and are coeval with some of those pre-Upper Jurassic extrusives and sediments found in the Insular belt of Monger, et al. (1972), which includes Vancouver Island.

The Island Copper porphyry deposit is associated with a stock and dikes within a major shear zone. Movement on this shear appears, based on regional geology, to be right lateral and sulfide mineralization occurs in an elongate pattern within the shear beyond the Island Copper deposit. Superficially, at least, it appears that intrusion, mineralization, and displacement on the dextral fault were overlapping. This deposit, with a 153 m.y. date, would seem to have formed during a period of San Andreas type fault activity

during which tectonic transport of the allochthon occurred.

In the post 80 m.y. period strike-slip tectonics were again dominant following relaxation of strong regional compression (Monger, et al., 1972). As noted previously, however, where Tertiary porphyry deposits developed in Lower Mesozoic volcanic terrane they formed as stock-type porphyry copper deposits. The type of crust penetrated by the mineralized intrusion apparently exerted significant control over its metallogeny. The absence of rift tectonics in the Tertiary coincides with an absence of Tertiary diorite porphyry copper deposits. The diorite model occurs only in association with persistent normal faults of the Lower Mesozoic.

Fig. 33 shows the location of porphyry molybdenum (after Clark, 1972) as well as porphyry copper deposits in such a manner that their spatial distribution may be appreciated.

TYPES OF PORPHYRY DEPOSITS IN THE CANADIAN CORDILLERA

Different crustal settings have been accompanied by three distinct types of porphyry copper deposits. Examples of the diorite, granitic pluton, and stock types are described in that order.

The Diorite Model

Definition of Diorite Model Porphyry Copper Deposits: The diorite model is defined as a quartz-deficient, commonly zoned, diorite-monzonite-syenite pluton with pervasive alteration and disseminated sulfide copper occurring in or concordant with the diorite, monzonite, or syenite phase. Petrographic zoning may be absent so a diorite could host the sulfides alone; if so, the diorite may be alkalic or calc-alkalic. Diabase may provide a similar response to the action of hydrothermal fluids, as may extrusive equivalents of the diorite-monzonite-syenite suite, so these rocks are included in the model.

Petrographic variations from diorite to syenite of the rocks that host individual diorite model porphyry copper deposits lead to great differences in silicate alteration mineral assemblages within the model. Generalizations concerning petrography, alteration, and mineralization are made and followed by specific examples of soda-potash (Ingerbelle-Copper Mountain), sodic (Afton), and potassic (Galore Creek) deposits. Diorite model deposits have been found in the Pennsylvanian-Permian-Triassic marine volcanic formations of Alaska but none has yet proven commercial. These deposits therefore could occur from Alaska to the Idaho batholith, where upper Paleozoic-Lower Mesozoic marine volcanic rocks occur. Most importantly, however, the diorite porphyry copper deposits occur within three elongate belts in British Columbia (Fox, 1975) within the Nicola-Takla-Hazelton formations. All examples cited in this chapter are from British Columbia.

Petrography and Alteration in the Diorite Model: Sutherland Brown, et al. (1971) and Field, et al. (1974) indicate that the diorite model actually includes syenites, monzonites, alkalic gabbro, and other quartz-deficient intrusions in addition to diorite. Most phases in the intrusive sequence are typically porphyritic. Fox (1975) encourages the speculation that most dioritic rocks are part of either nepheline or leucite normative alkalic magma series distinct from the calc-alkaline suite that gives rise to quartz-bearing intrusions, a discussion further refined in *CIM Special Volume 15* (1976). This petrochemistry is summarized in the ternary diagram of Fig. 36b.

Undersaturated intrusions other than diorite could derive from a dioritic parent by potassic metasomatism, through processes of assimilation or differentiation, or some combination of these. However, the occasional presence of magmatic(?) garnet, nepheline, analcite, leucite, and other minerals not characteristic of calc-alkaline intrusions supports the hypothesis that they are a distinct magmatic series. Similarly, the time relationship between early diorite and late syenite in each zoned porphyry copper-bearing magmatic sequence suggests the two are genetically associated in composite zoned differentiated intrusions.

The unifying feature of the quartz-deficient intrusion that hosts porphyry copper mineralization regardless of magmatic composition is the fairly uniform response of the rock to sodium, potassium, and hydrogen metasomatism. In most deposits a potassic zone is surrounded by a chlorite-dominant propylitic zone. If sodium metasomatism developed instead of potassium the potassic zone may not appear. The mineral assemblages in the propylitic zone may vary widely in detail from deposit to deposit, but if a potassic zone has developed it is always dominated by chlorite- or biotite-rich assemblages or both. In only a few deposits is a phyllic or an argillic zone developed. Rather, a central potassic zone tends to pass directly to the outer chlorite-rich propylitic zone with no intermediate stages developing, in contrast to the silica-rich porphyry copper deposits (quartz monzonites) described by the Lowell and Guilbert (1970) model of potassic-phyllic-argillic-propylitic zoning from the core outward. The pyrite halo in the diorite model may be greatly diminished in both volume and intensity compared to the more silicic model.

With only two alteration zones generally discernable in diorite deposits within the terms of the Lowell and Guilbert (1970) alteration zone definitions, copper is less uniformly fixed in any one zone. It may occur in what is mineralogically a propylitic zone, as at Duckling Creek, or in the potassic zone, as at Galore Creek. Metallization may also occur in both potassic and propylitic zones or in the phyllic zone in the rare instances of its presence. This results in a weakened reliance on zoning as an exploration tool in the search for a copper center, which usually includes the potassic zone if present but is not restricted to this.

Sodium metasomatism has been reported at Afton (Iron Mask batholith) and Ingerbelle (Copper Mountain stock) but mineralogy of this alteration suite has not been well documented. The Copper Mountain stock contains a porphyry copper deposit associated with a potassic center (the Copper Mountain ore body) as well as one associated with a sodium metasomatism (the Ingerbelle ore body). Sodium metasomatism has also been recognized by Hollister, et al. (1975),

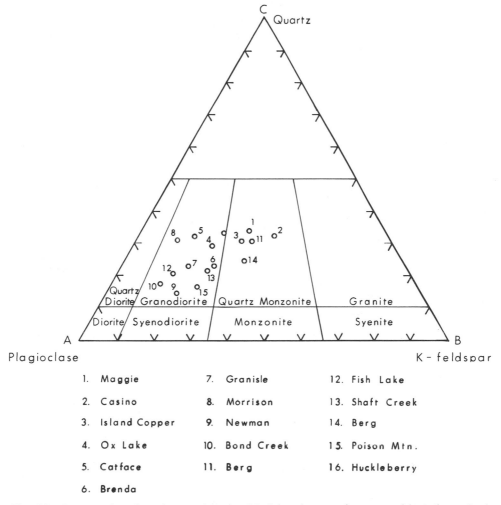

Fig. 36a. Petrography of stock type deposits. Modal and normative compositions from fresh rock in porphyry complexes where available in the literature are shown in this figure. Because fresh rock analyses are used the composition given is not that ordinarily found spatially and temporally close to ore and composition does not reflect the phase that accompanies ore (*modified after Carter, 1974*).

1. Maggie	7. Granisle	12. Fish Lake
2. Casino	8. Morrison	13. Shaft Creek
3. Island Copper	9. Newman	14. Berg
4. Ox Lake	10. Bond Creek	15. Poison Mtn.
5. Catface	11. Berg	16. Huckleberry
6. Brenda		

in diorite model porphyry deposits in Alaska.

Mineralization in the Diorite Model: Mineralization in diorite model porphyry copper deposits tends to occur most commonly in spatial association with monzonite or syenite phases of zoned diorite-monzonite-syenite intrusions. Copper sulfide is accompanied only rarely by molybdenum but commonly by abnormally high gold:copper ratios. Sulfur accompanying the copper is insufficient to consume the iron present in the host intrusion, and the altering mineralizing process leaves large amounts of magnetite accompanying chalcopyrite. Magnetite may be as important a hydrothermal mineral in dissemination in diorite model porphyries as is pyrite in quartz monzonite model deposits, occurring as a dissemination in the potassic and propylitic zones. Mineralizing fluids appear to be silica-deficient because

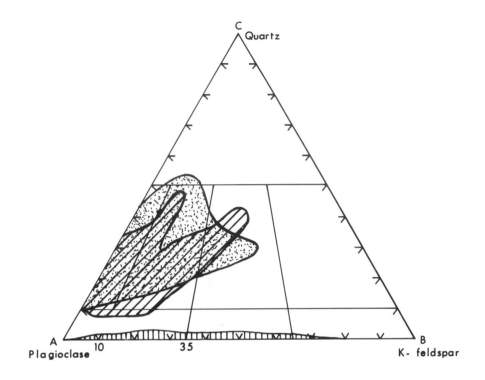

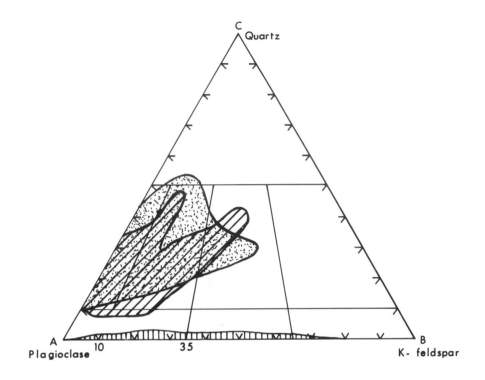 Area including modal compositional trends for diorite model intrusions associated with northern Cordilleran porphyry copper deposits (Compiled from data in CIM Spec. Vol. 15, 1976).

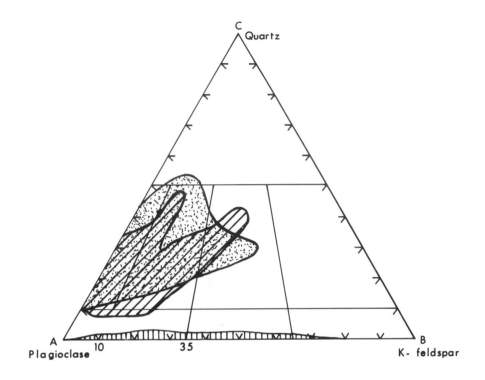 Area including modal compositional trends of the Guichon Batholith (Highland Valley) (Compiled from data by Northcote, 1969).

 Area including modal compositional trends of Cascade mineralized plutons (Compiled from data from various sources).

Fig. 36b. Petrography of diorite, granitic pluton, and Cascade deposits. Compositional trends for plutons associated with diorite model porphyry copper deposits are compared with trends for plutons associated with continental margin deposits and the Guichon batholith. The Cascade trends are believed typical of continental margin belts underlain by oceanic crust and are similar to trends established for the Antilles.

veinlets containing the ore sulfide may be free of quartz but may contain epidote, chlorite, calcite, prehnite, or zeolite.

Stockworks are much more common in diorite model porphyry copper deposits than breccias. Volumetrically, dissemination of ore minerals in the potassic zone is important in most intrusions, while fracture filling is most important in zones adjacent to the potassic.

Fox (1975) cites examples of the diorite type porphyry deposits where a central biotite zone carrying chalcopyrite-bornite-magnetite or bornite-magnetite gives way to a peripheral zone of chlorite-epidote-albite with associated pyrite-chalcopyrite. Gold is concentrated in the bornite-magnetite zone in some deposits.

Zinc and lead occurrences attributed to zoning are rare in a copper-zinc-lead sequence around the copper heart of diorite deposits. The diorite model does not seem to occur in a zoned district.

Ingerbelle-Copper Mountain: A Diorite Type Example: The Ingerbelle-Copper Mountain complex is a fairly typical diorite model deposit. Since it was discovered in 1884 and was the first porphyry copper deposit mined in British Columbia, a substantial bibliography exists describing its geologic features. This summary, however, emphasizes selected features from the description given by Ney and Brown (1972). The Ingerbelle ore body has a reserve of 69 million mt (76 million tons) of 0.53% Cu.

In the Ingerbelle-Copper Mountain area a complex zoned diorite-monzonite-syenite pluton with a radiometric date of about 193 m.y. intrudes Nicola group (map unit three in Fig. 32), which is composed of fragmental marine andesites, water-laid tuffs, and volcanic siltstones. Except in the immediate vicinity of the plutonic rocks, deformation, metamorphism, and metasomatism are very mild.

The intrusions found in and near ore are quartz-poor, porphyritic syenite in composition, and albitized. They are popularly regarded as genetically related to the Copper Mountain stock through differentiation, although surface outcrop is inadequate to clearly establish this. The close spatial relationship between igneous rocks and ore has led to the common assumption that a genetic relationship also exists.

Mineralization has been accompanied by potassium, sodium, hydrogen, and sulfur metasomatism. Within 60 m of the Copper Mountain stock, hydrothermal alteration has overprinted a granoblastic development of diopsidic pyroxene, hornblende, biotite, epidote, and intermediate plagioclase. Hydrothermal alteration has developed in several successive stages. Development of a biotite

rich but orthoclase-poor potassic zone appears to have been the earliest stage but this has been largely obliterated by successive hydrothermal events. The early potassic zone was probably widespread as remnants are found at widely separated localities. Magnetite formed intergrowths with the biotite. Orthoclase is common only near Copper Mountain. The remaining biotite zone includes the Copper Mountain ore body. Within the potassic zone, an assemblage of albite, epidote, sphene, apatite, and minor pyroxene succeeded development of pervasive biotite along some fractures and areas of fracturing. Although not district-wide in its appearance, this type of alteration has converted some andesite to megascopically appearing diorite-textured andesite.

Scapolite, albite, calcite, chlorite, and other hydrated iron silicates make up the last stage of alteration. The scapolite-chlorite assemblage replaced part of the biotite zone as well as apparently fresh rock. It surrounds the potassic zone and is popularly considered equivalent to the propylitic zone. The scapolite-chlorite assemblage includes the Ingerbelle ore body.

Within the Ingerbelle ore body albite, chlorite, and carbonate are more important than quartz as gangue occurring with ore sulfides.

From the sequences noted it is clear that a quartz-sericite dominant phyllic zone is not important in this area. Hypogene copper sulfide mineralization is found in what was the early potassic alteration phase as well as in alteration assemblages peripheral to and replacing it. If the albite-scapolite alteration assemblages may be termed propylitic, then the propylitic zone occurs adjacent to and probably replaced part of the potassic and contains significant hypogene copper sulfide. Within these definitions the Copper Mountain ore body was found largely in the potassic alteration zone, whereas the Ingerbelle occurs largely in the propylitic.

Afton: A Sodium-Rich Diorite Model Example: Afton has been described by Carr (1976) as a steeply dipping 27 million mt (30 million ton) ore body containing 1% Cu that occurs at the west side of the nepheline-normative 198 m.y. Iron Mask batholith. This is a zoned pluton with diorite, monzonite, and syenite stages, with ore occurring near a syenite outcrop. The ore minerals show a vertical and lateral concentric zoning outward from a core containing native copper through chalcocite to bornite, then chalcopyrite, and finally to pyrite outer and lower zones. The extent of iron oxidation increases markedly with increasing alteration and accompanies the economically important copper mineralization, leading Carr (1976) to conclude that the native copper formed as a result of supergene processes.

Alteration minerals accompanying the different copper zones suggest extensive aluminum and sodium metasomatism. The zonal pattern of the gangue and ore minerals grades outward from a core with goethite-prehnite-albite-chlorite-chalcocite-native copper downward and laterally through albite-epidote-chlorite-hematite-bornite-chalcocite, albite-chalcopyrite-bornite-chlorite, to an erratic outer pyrite-magnetite bearing phase. A chlorite-rich propylitic zone surrounds the pyrite-bearing halo.

Quartz and molybdenite are notable for the rarity with which they are found to fill fractures.

The inferred composition of the ore fluid based on the mineralogy present suggests influence by alkalic undersaturated rocks.

The ore occurs in veinlets and as a cement for breccia of altered host rocks. The suggestion implied by repetitive veining and brecciation is that openings that reappeared over a considerable time span became filled with vein matter deposited at successively lower temperatures. Alteration mineral zoning appears compatible with that described in the McCarthy and Jacobsen (1976) model.

Galore Creek: A Potassium-Rich Diorite Model Example: The Galore Creek deposits described by Barr (1966) are ten hypogene cupriferous sulfide concentrations localized in highly fractured and altered Triassic intrusive and extrusive rocks. The intrusions are parts of a potassium-rich syenite complex, and the deposits, which occur over an area 3.2 x 5.6 km (2 x 3.5 miles) may be separated into either porphyry copper or pyrometasomatic types.

The syenite consists of a number of phases rich in orthoclase and poor in mafic minerals. None contain as much as 5% modal quartz. Feldspathoids have been identified in some units.

Both extrusive fragmental and cataclastic breccias occur within porphyry copper deposits in the complex. In areas where potassium metasomatism has not been too severe breccia fragments are clearly discernible and the character of fragmentation permits classification as to type. However, contacts between various units of the complex are frequently masked by intense alteration and in most cases the origin of a breccia cannot be distinguished with any degree of certainty. Sulfide occurs as a cementing mineral in breccias clearly of cataclastic origin.

Within copper sulfide-bearing areas the most consistent and directly asociated secondary silicate encountered is biotite, which generally occurs as a fine-grained felted mass and may make up 50% of the rock. Most mineralized zones also contain orthoclase as veinlets and massive replacements commonly associated with biotite where sulfide is present. Anhydrite, garnet, magnetite, and apatite may occur in this potassic zone but without a consistent spatial relationship to sulfide. Pyrite is pervasive in the potassic zone. All major units of the complex are propylitized if potassic (biotite-orthoclase) zone alteration is not present. Epidote, chlorite, and lesser sericite replaced the original magmatic silicates in the propylitic zone. Alteration zoning therefore consists of potassic-propylitic with copper strongest in the potassic core.

Quartz Monzonite Model Granitic Pluton Type

Granitic pluton type porphyry copper deposits (after Field, et al., 1974) occur associated with zoned quartz diorite-granodiorite-quartz monzonite intrusions of batholith size, with porphyry deposits frequently occurring in the interior or core associated with the youngest phase. These plutons occur from the Cuddy Mountain-Peck Mountain area in Idaho, as noted by Field, et al. (1974), to the Gibraltar deposit in British Columbia. All are intrusive into the Triassic-Jurassic Takla-Hazelton-Nicola volcanic assemblage or its equivalents in Idaho and Washington and commonly are comagmatic with their host. With the possible exception of Brenda with its inconclusive K-Ar date, no granitic pluton deposits have been found with dates appreciably younger than the intruded calc-alkalic volcanic rocks. All appear to occur in the 205-170 m.y. range except Brenda. Table 12 lists the largest known ore deposits of this type.

Petrography in the Granitic Pluton Type: The granitic plutons that host porphyry copper deposits are upper mesozonal to epizonal multistage intrusions containing concentric phases with early, more mafic nonporphyritic and sheared quartz diorite margins and younger more silicic, alkaline, and porphyritic cores. The most common magmatic terminal phase is close to quartz monzonite in composition. Petrography of this type is shown in the ternary diagram of Fig. 36b.

Northcote (1969), Ager, et al. (1972), and Hylands (1972) make an excellent case for magmatic differentiation in the Guichon batholith and their petrographic interpretations are applicable to other plutons (e.g., Cuddy Mountain, Gibraltar) similar to the Guichon. Their strongest evidence is the gradational contact that may be seen between any two adjacent petrographic stages. Acceptance of this hypothesis permits speculation that mineralization may be an end product of differentiation of a calc-alkalic magma and the evidence is continuous enough to be plausible. The most effective process of differentiation appears to have been inward residual enrichment of alkali and silica in the crystallizing magma.

Structure and Mineralization: Sulfide regularly occurs in the core zone closely allied with the youngest intrusive phase, adding support to the differentiation-mineralization concept wherein as the core unit crystallized, confining pressure was exceeded by increasing internal pressure, and hydrothermal

fluids escaped in silicate-silica-sulfide breccia pipes, dikes, and stockworks. Existence of late dikes occurring with stockwork copper-molybdenum deposits suggests a common origin and emplacement control for both. In the individual ore bodies, stockworks that trend with the major faults cutting the batholith dominate.

Hollister, et al. (1975) suggest a structural control at Highland Valley for evolution of the pluton (from early quartz diorite to late quartz monzonite), alteration from early potassic to late phyllic, and mineralization (beginning comagmatic with late granodioritic intrusions subjected to potassic alteration and ending with late quartz monzonite intrusions, some of which have phyllic alteration associated with ore). The timing of events based on their fault reconstructions provides continuous link between wholly magmatic events and magmatic events that accompany hydrothermal activity. Their fault reconstructions interpret the mineralization stages to coincide with magma ranging from granodioritic to quartz monzonitic in composition. In this reconstruction the ore bodies that appeared early are clothed in a potassic alteration suite, whereas the later ones occur in an extensive phyllic zone. Fig. 37 illustrates the structural developments at Highland Valley.

Drummond, et al. (1972), show that sulfide mineralization occurs surrounding the quartz-feldspar porphyry core of a zoned intrusion at Gibraltar. Their evidence is compatible with derivation of the sulfide from a large differentiating calc-alkalic pluton.

Soregaroli (1971) describes mineralization at Brenda in what he calls a zoned and composite intrusion. The general setting appears similar to that found in the Guichon, although Brenda mineralization occurs at the point of contact between the pluton and the intruded rock.

Table 12 lists the alteration zoning sequence from the center outward. All samples cited have a well developed potassic zone and most have well defined phyllic zones. An argillic zone developed in only a few High-

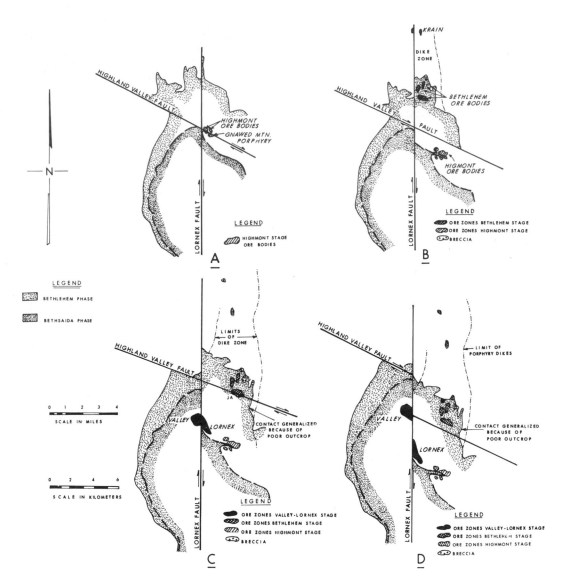

Fig. 37. Structural evolution of the Highland Valley. Movements on the Lornex and Highland Valley faults occurred simultaneously and alternatively in the final phases of intrusion of the Guichon batholith. The fault planes provided the openings for the admission and deposition of mineral and igneous matter. Displacements on these two major faults permitted a dilation of the host rocks by an order of 20%, as indicated by the amount of dikes and vein material now present that constitute the ore bodies. The earliest stage of hydrothermal activity appears to have been in and close to the Highmont ore body, which has the lowest copper-molybdenum ratio in the district. This ore body **(Fig. 37A)** formed as soon as consolidation of the differentiating pluton had advanced to the point where walls were strong enough to support openings at and near the junction of the two faults. **Fig. 37B** shows the configuration of the batholith at the end of the formation of the Bethlehem ore bodies. Displacement of various fault blocks infers simultaneous confined movement on both the Lornex and Highland Valley faults from the time of Highmont mineralization through the development of all Bethlehem ore bodies. The Bethlehem ore bodies formed in

land Valley ore deposits (i.e., Lornex deposit) but is missing in the other examples. Pyrite zones around this type of porphyry copper deposit are usually erratically developed—i.e., they do not extend far beyond the ore body, do not ordinarily contain more than 3% pyrite, and the area containing disseminated pyrite ordinarily includes large areas of rock barren of pyrite.

A substantial percentage of the ore sulfides in potassic zones are disseminations in the granitic pluton type, whereas fracture filling of sulfides clearly dominates in those zones adjacent to the potassic. No significant copper or molybdenum sulfides are found outside the argillic zone.

Copper may occur in both the potassic zone, characterzied by orthoclase-biotite or orthoclase-chlorite mixtures or both, and in the phyllic zone, essentially a quartz-sericite-pyrite mixture. Chalcopyrite is the dominant copper mineral and copper sulfosalts are rare in most deposits. All ore zones carry some molybdenum. Zinc and lead may be zonally arranged about the copper center.

Quartz Monzonite Model Stock Type

The stock type porphyry in the area of interest is a mineralized boss, dike zone, or stock of calc-alkalic composition injected into its host rock. The mineralized pluton may be quartz diorite to quartz monzonite in composition, but quartz monzonite is the most common igneous stage intruded close to ore in time.

Burchfiel and Davis (1973) point out that igneous intrusions of this type may be structurally anologous to salt plugs intrusive into sediments. Gussow (1962) reaches a similar conclusion for igneous masses. The different structural types suggested by Sutherland Brown (1972) and the examples given by Carson and Jambor (1974) in the Babine Lake area for porphyry copper deposits associated with smaller intrusions suggest that in many places the magma is an igneous piercement or diapir intrusive into wall rock that yielded as injection took place.

Distribution of porphyry copper prospects having a potential in excess of 18 million mt (20 million tons) of 0.1% and displaying porphyry type alteration is shown in Fig. 35. Most fall within the stock type classification. As noted in Table 13, most intrude the Triassic-Jurassic volcanic rocks, although most postdate these extrusives and are innocent of the tectonic environment present during that older period of volcanism.

The majority of stock type porphyries are under 83 m.y.; only a few (e.g., Island Copper) are older. The tendency of these mineralized intrusions to occur during periods of time when formation of batholiths was minimal suggests that they may have a magmatic origin distinct from that which gave rise to the large intrusive masses associated with rapid subduction (e.g., Sierra Nevada type batholiths). This negative correlation also suggests that each may occur under differing conditions of plate movement, with batholiths more easily identified with rapid subduction and the diapiric model more closely associated with strike-slip tectonics.

Petrography of the Stock Type: The intrusion most intimately associated with porphyry copper mineralization of the stock type is commonly close to quartz monzonite in composition. Most such intrusions are mixtures of quartz, plagioclase, orthoclase, biotite, and hornblende. All known well mineralized intrusions are porphyritic. Fig. 36a (Carter, 1974) shows petrographic trends of these plutons. Magmatic quartz and ortho-

---

the various strands of the Lornex fault. Sulfide and gangue silicates occupied these strands as they formed. The results are the breccia and veinlets that make up the ore body. Dike intrusion accompanied mineralization and the ultimate limit of porphyry dikes is shown in Fig. 37D. Fig. 37C shows the position of the batholith at the time of Valley-Lornex mineralization, which followed Bethlehem-stage ore deposition. Fig. 37D shows present configuration of the ore bodies with the Lornex displaced from the Valley by postmineral movement.

Potassic alteration suite minerals dominate in both the Highmont and Bethlehem stages but sericite-rich phyllic is more widespread in the Valley-Lornex stages, suggesting a change in chemistry of the hydrothermal fluids. The Valley has among the lowest molybdenum: copper ratios in the district. The relationship of ore sulfides to dikes indicates that both may have been derived from a differentiating magma. Displaced fault strands are not shown in these figures (*after Hollister and others, 1975*).

clase may be restricted to the groundmass, as at Granisle, or may occur as prominent phenocrysts, as at Berg.

Other intrusive stages may exist in addition to the quartz monzonite, but composition varies from one intrusive center to another. The other igneous stages usually preceded the quartz monzonite with its closely associated mineralization.

Mineralization and Structure: Mineralization could have accompanied, followed, or been active both during and after injection of the igneous rock. The structure that developed as a result of injection may have controlled or dominated contemporaneous or subsequent entrance of hydrothermal fluids. Breccias commonly are developed with this type of porphyry copper. On the other hand, if the stock is intruded into a regional fault zone or line of weakness a stockwork may develop that reflects this governing structure. The principal veinlet trend parallels that of the major fault. Mineralization is always included with later steps in the sequence of igneous and hydrothermal events.

The injection of a stock also frequently results in development of a circular fracture pattern. The fractures may be mineralized to form an annular stockwork containing a major veinlet trend parallel to the intrusion contact.

Alteration zones may not be as simply developed in the stock type as in the granitic pluton. In the case of porphyry copper deposits associated with small injected intrusive bodies, the hydrothermal fluid may have been a lubricant since rocks near the contact of the pluton usually show stronger hydrothermal metasomatism than do the central parts. Should the intrusive contact be the locus of a potassic zone, the phyllic zone may occur both internally in the stock and externally away from it. On the other hand, if the center of alteration is well within the stock the phyllic, argillic, and propylitic zones will occur outside the potassic in concentric shells in either the pluton or the intruded rock or both.

Copper and molybdenum sulfides commonly are best developed in the contact zone. If this is also the potassic zone, dissemination rather than fracture control of values may be significant. Hypogene values weaken away from the contact zone, although the stock itself frequently contains low grade mineralization. Away from the potassic zone the ore sulfides fill either space between breccia fragments or tectonic voids that may be related to either the force of the intrusion or movement on a regional structure. Sulfide mineral zoning also is frequently related to the contact the pluton makes with the intruded rocks. Molybdenum in some deposits is more conspicuous near the contact than in the walls.

A number of deposits have scattered zinc and lead veins occurring outside the copper zone, suggesting that metallogenic zoning may develop around the stock.

This type of porphyry is most commonly copper-molybdenum with relatively low gold values, although copper-gold types are also known.

The Berg: An Example of the Stock Type: The Berg deposit in west central British Columbia has been the subject of numerous short contributions from a number of sources. This geologic outline summarizes pertinent data from the large bibliography available on the deposit, primarily following Sutherland Brown, et al. (1971).

The Berg composite stock is a cylindrical pluton 800 m in diameter with a 50 m.y. K-Ar date. Composition ranges from an older quartz diorite to a younger quartz monzonite porphyry phase. The pluton intrudes the Hazelton group (map unit three in Fig. 32).

Mineralization and alteration are strongest at the contact between the stock and the intruded rock. The annular potassic zone found in and near the contact contains topaz as well as secondary biotite and orthoclase. Hypogene copper and molybdenum sulfides are zoned in and near this potassic zone with a molybdenite zone occurring closest to the stock contact and copper occurring as an annular zone outside it. A thin discontinuous sericite zone rings the potassic zone and both give way to a pyrite-bearing chlorite-rich propylitic zone developed in the Hazelton rocks. Hypogene copper values reach their peak at

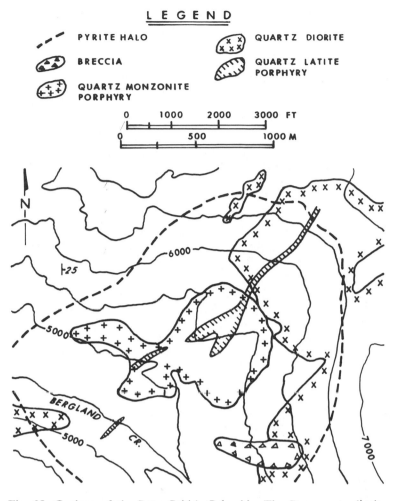

LEGEND

- - - PYRITE HALO

BRECCIA

QUARTZ MONZONITE PORPHYRY

QUARTZ DIORITE

QUARTZ LATITE PORPHYRY

Fig. 38. Geology of the Berg, British Columbia. The Berg quartz diorite-quartz monzonite-quartz latite stock, where sulfide is best developed along and near the margin of the quartz monzonite, is typical of many stock-type deposits. Best grade hypogene sulfide mineralization is found near the north and west contacts of the quartz monzonite (*after Carter, 1974, and Sutherland Brown, et al., 1971*).

the approximate interface of the biotite and chlorite-dominant zones.

The Babine Lake Deposits: Examples of the Stock Type: Carson and Jambor (1974), Carter (1974), and Carter and Kirkham (1969) provide excellent background for summarizing the numerous stock-related porphyry deposits found in and near Tertiary quartz-bearing calc-alkalic plutons acting as diapirs in the Babine Lake area. These stocks are about 51 m.y. in age, intruding rocks of map unit three (Triassic and Jurassic volcanic rocks).

Within the pluton at Bell and Granisle an amphibole zone may be interpreted as the core of alteration and mineralization. Carson and Jambor (1974) point out that hydrothermal amphibole could be paragenetically later than secondary biotite. However, diminishment of copper in this zone coupled with enrichment of copper in the biotite zone (which may surround amphibole) argues that it is a core developed early in the mineralizing process. In mineralized plutons other than Bell and Granisle a central biotite or biotite-chlorite zone usually carries some secondary orthoclase in mineralized plutons where the amphibole zone is missing. Sodic plagioclase is intergrown with orthoclase. Some Babine Lake deposits show a transition from an inner biotite zone to a quartz-sericite-pyrite assemblage (as at Bell and Granisle), whereas some smaller deposits show a direct transition from the biotite to a chlorite-carbonate zone.

In the larger deposits bornite-magnetite or bornite-chalcopyrite mineralization in the potassic zone gives way to chalcopyrite-pyrite and then pyrite as distance is gained from the center of mineralization. Molybdenum grades less than 0.01% Mo and gold is not usually prominent.

Conclusions for Deposits in the Canadian Cordillera

Mineralization in the porphyry copper province of the northern Cordilleran orogen began in the Upper Triassic with what appears to have been a rifting environment. Diorite model and granitic pluton type porphyry copper deposits formed in the Triassic and Lower Jurassic as the trough was filled with volcanic debris. This was followed by a relatively sterile period in the Upper Jurassic and Lower Cretaceous when plutonic activity was batholithic in nature. The Upper Cretaceous and younger periods witnessed a decline in batholithic intrusive activity and a resurgence of smaller stock type porphyry copper-related intrusions. These later deposits tend to be related to circular or oval plutons that may bear a tectonic resemblance to diapirs. They also coincide with development of a strike-slip tectonic regime that became resurgent at the end of the Cretaceous.

What have been called quartz monzonite or calc-alkalic types in this chapter could be incorporated into the Lowell and Guilbert (1970) model if that model were expanded to include the large (e.g., Highland Valley) and small (e.g., Berg) differentiated deposits and if the aberrations in alteration characteristics listed in Tables 12 and 13 also could be accommodated by that model.

The very large granitic pluton types are usually old (Triassic to Middle Jurassic) porphyry copper deposits that have been deeply eroded. Consequently, their alteration zoning and geometric pattern may not fit the typical Lowell and Guilbert (1970) model. Their outlines tend to be governed by the position of copper mineralization in a differentiating pluton of batholithic size. In the case of Highland Valley the aggregate reserves of copper-bearing rock now known exceed 4000 mt and the individual ore zones (e.g., Valley, Lornex, and Bethlehem) have been called fault displacements, as the shapes present are markedly different from the ovals of the Lowell and Guilbert (1970) model.

Such stock type deposits as Berg represent a porphyry copper morphological type that also should be incorporated into the Lowell and Guilbert (1970) model. Mineralization and alteration of the stock type are usually strongest in the vicinity of a stock contact. The resulting doughnut-shaped zone of sulfide still should be considered part of the Lowell and Guilbert (1970) model since petrography, ore mineralogy, and alteration are typical.

Deep erosion at some deposits (e.g., Highland Valley, where the present surface may have been 5 to 6 km below the surface at time of mineralization) may also expose alteration zones that appear to be aberrant from the Lowell and Guilbert (1970) model. The Bethlehem ore bodies, for example, do not have exposed argillic or phyllic alteration assemblages accompanying the potassic zone with its ore minerals. Such zones could have occurred much higher in the system and their absence from some granitic pluton type ore bodies may be a function of depth of erosion alone. The Lowell and Guilbert (1970) model therefore should be modified to include the quartz monzonite (calc-alkalic) model of the Canadian Cordillera.

At the Berg and other similar deposits existence of metallization at and near the contact could be used as evidence for scavenging of metals from intruded wall rock in a meteoric-hydrothermal system generated by heat from a stock. On the other hand, presence of copper-bearing potassic zones within the stocks (e.g., Babine Lake and Highland Valley areas) suggests the possibility that metals were brought in by intrusion. The paucity of molybdenum in the Babine Lake deposits is compatible with their setting (basic marine andesitic volcanic rocks).

PORPHYRY COPPER DEPOSITS OF THE CASCADES

Introduction

Within the Cascades of the northern Cordilleran orogen is a porphyry copper province distinctive in mineralogy, metallogeny, and time distribution. Because of the differences distinguishing these deposits, the Cascades should be treated as a separate province. This section briefly examines characteristics of this province and offers explanations for the differences observed.

The area encompassed by this study extends from 46° to 49° north and is confined to a northeast-trending belt of mineralized plutons found west of the Pasayten fault confined to the area from 120° to 123° west, or an area extending approximately from Yakima to Seattle, WA. East of this province lie the Miocene Columbia River basalts; to the west lies the Tertiary sedimentary and volcanic complex that includes the Olympics. Fig. 39 shows the area covered and Table 14 lists the characteristics of porphyry copper occurrences in this area described in the literature. Although porphyry copper occurrences are common in the Cascades, none of the deposits yet explored are large. Because Table 14 compiles only selected published data, it clearly is not a complete listing of all porphyry copper occurrences in this region. Other deposits are known (e.g., Tabor, 1963) but the compilation in Table 14 adequately summarizes the characteristics of Cascade occurrences.

Geologic Setting of the Cascades

Fig. 39 presents those geologic elements of the Cascades most pertinent to development of this Tertiary porphyry copper province. Thrust faults commonly are not associated with porphyry copper deposits and to avoid cluttering this figure numerous thrusts reported by Misch (1966) are omitted, as are the many formation names. Strike-slip faults and plutons are shown in Fig. 39, however, because these features are closely associated spatially with porphyry copper deposits.

General Geology: Most porphyry copper deposits in this area are Tertiary but their characteristics reflect their setting. The best summaries of geologic setting yet published are by Misch (1966) and McKee (1972). Mattinson (1972) gives excellent radiometric data, while Hutting (1956), Purdy (1954), and Culver and Broughton (1945) provide pertinent data on mineral deposits. Grant (1969) summarized much known data on mineral deposits of the Cascades. This section on general geology selectively abstracts pertinent data important to the development of porphyry copper deposits from these authors. For additional details, the reader should study the references cited.

Crustal Composition: Southwest of the Helena and Deception Creek faults (Gaultieri, et al., 1973, and Grant, 1969) most exposures are Tertiary sedimentary and volcanic rocks with only minor Mesozoic material appearing at the surface. Preponderance of a Tertiary sedimentary and volcanic terrane in this area effectively masks pre-Tertiary crustal evolution, and little may be said at this point about host rocks penetrated by porphyry copper deposits south of these faults.

Oceanic crust is suspected to be a basement underlying a large part of this area (King, 1969). Hill (1975) interprets seismic data to suggest that the volcanic Cascades may not have a significant crustal root.

Between the Helena and Straight Creek faults the most common exposures are of Chilliwack group Devonian to Permian basic marine volcanic rocks, their differentiation products, and their associated volcanogenic sediments. If these rocks are equivalent to part of the Sicker group of the same age and lithology on Vancouver Island, it is conceivable that right lateral movement on the Deception Creek fault and its branches to the northwest may have been about 161 km (100 miles). Most displacement in this speculated correlation between the Sicker and the Chilliwack groups would have to have been pre-Middle Tertiary, as the various Ter-

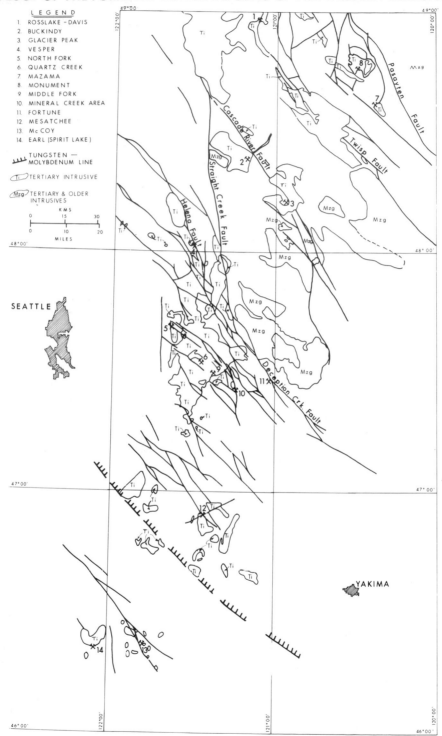

Fig. 39. Structural setting of porphyry copper deposits of the Cascades. Shear couples interpreted for Fig. 33, the compilation of structure of the northern Cordilleran orogen, apply equally to the compilation of strike-slip faults of the Cascades. Thrusts, fold axes, and drainage are omitted to simplify the diagram (*compiled from various US Geological Survey maps*).

Table 14. Tertiary Porphyry Copper Occurrences of the Cascades

Map No.	Name in Literature	Location	Type	Pluton Age m.y.	Host Rock	Alteration Zoning Sequence from Center	Size, m x 10³ (10³ x ft)	Pyrite Zone Content in Phyllic Zone (%)	Structure Model	Major Fracture Trends	Metals Present	References
1	Ross Lake-Davis	121°08'; 48°58'	Qtz Dio Por	30	Tert Vol	Pot-Phy-Arg-Prop	1.2 x 1.2 (4 x 4)	4	Stockwork	NE-EW	Cu, Mo	Grant, 1969
2	Buckindy	121°11'; 48°22'	Grdr Por	—	Pal Gn	Pot-Phy-Arg-Prop	1.8 x 1.5 (6 x 5)	3	Stockwork	EW-NE	Cu, Mo, W	Grant, 1969
3	Glacier Peak	120°56'; 48°12'	Qtz Mon Por	22(?)	Pal Gn	Pot-Phy-Arg-Prop	2.1 x 1.5 (7 x 5)	5	Stockwork	EW-NE	Cu, Mo, W	Grant, 1969
4	Vesper	121°31'; 48°02'	Grdr Por	32	Pal Gn	Pot-Phy-Arg-Prop	1.2 x 1.2 (4 x 4)	3	Breccia	—	Cu, Mo, W	Grant, 1969
5	North Fork	121°37'; 47°37'	Qtz Dio Por	9.9	Tert Sed	Pot-Phy-Prop	—	Po	Stockwork	NW-NE	Cu	Patton, et al., 1973
6	Quartz Creek	121°29'; 47°35'	Qtz Mon Por	18	Tert Sed	Pot-Prop	1.2 x 0.9 (4 x 3)	—	Breccia	—	Cu, Mo, W	Grant, 1969
7	Mazama	120°22'; 48°36'	Qtz Dio Por	70	Cret Sed	Pot-Phy-Arg-Prep	1.5 x 1.5 (5 x 5)	3	Stockwork	—	Cu, Mo	Huntting, 1956
8	Monument	120°29'; 48°46'	Qtz Mon Por	49	Cret Sed	Pot-Phy-Arg-Prop	1.5 x 1.5 (5 x 5)	4	Stockwork	—	Cu, Mo, W	Eaton and Staatz, 1971
9	Middle Fork	121°22'; 47°29'	Qtz Mon Por	18	Tert Sed	Pot-Phy-Prop	4.6 x 1.2 (15 x 4)	4	Stockwork	NW-NE	Cu, Mo	Grant, 1969
10	Mineral Creek	121°15'; 47°25'	Grdr	Tert	Tert Sed	Pot-Phy-Prop	1.5 x 0.9 (5 x 3)	Po	Stockwork	EW-NW	Cu, Mo, W	Grant, 1969
11	Fortune	121°04'; 47°27'	Dac Por	Tert	Cret Sed	Pot-Phy-Arg-Prop	1.8 x 1.2 (6 x 4)	4	Stockwork	NW-NE	Cu, Mo, W	Gaultieri, et al., 1973
12	Mesatchee	121°24'; 47°50'	Dac Por	6.2	Tert Vol	Pot-Phy-Arg-Prop	—	Po	Stockwork	NW-NE	Cu, Mo, W	Simmons, et al., 1974
13	McCoy	121°47'; 46°22'	Qtz Dio Por	24	Tert Vol	Pot-Phy-Arg-Prop	1.2 x 1.2 (4 x 4)	Po	Stockwork	NW-NE	Cu	Huntting, 1956
14	Earl (Spirit Lake)	122°05'; 46°21'	Qtz Dio Por	16	Tert Vol	Pot-Prop (Tour)	1.8 x 1.2 (6 x 4)	—	Stockwork	NW-NE	Cu	Huntting, 1956

Tert: Tertiary
Qtz: Quartz
Dio: Diorite
Por: Porphyry

Mon: Monzonite
Cret: Cretaceous
Sed: Sediments
Gn: Gneiss

Vol: Volcanics
Dac: Dacite
Grdr: Granodiorite
Pal: Paleozoic

Pot: Potassic
Phy: Phyllic
Arg: Argillic
Prop: Propylitic
Po: Pyrrhotite

tiary plutons do not appear to be offset by this fault system.

Pre-Devonian crystalline rocks outcrop between the Straight Creek and Helena faults, including the Yellow Astor gneiss with its Precambrian radiometric dates (Mattinson, 1972, and Misch, 1966). Although much of the exposed pre-Devonian consists of thrust slices overlying the Devonian-Permian Chilliwack group, allochthonous outcrop of these older rocks to the northwest favors existence of a pre-Devonian cratonic basement for this segment of the Cascades.

Between the Straight Creek and Deception Creek faults on the west and the Cascade River fault (Tabor, 1963) on the east, the crust consists of pre-Mesozoic gneisses invaded by Mesozoic and Tertiary plutons. Yellow Astor gneiss in this area includes Precambrian; however, the only clear dating for most of these rocks is unquestionably pre-Jurassic. However, most gneisses are believed to be metamorphosed Paleozoic formations. Whatever its age, the present metamorphic complex with large Mesozic and Tertiary batholiths occupying nearly half of the area simulates cratonic crustal conditions. The metamorphic terrane east of the Straight Creek fault is similar to another terrane to the north, on the west side of the northerly projection of this fault. The relationship has led to speculation that pre-Upper Cretaceous right lateral displacement on the Straight Creek fault is 193 km (120 miles) (Misch, 1966).

Gneissic rocks of pre-Upper Jurassic age are dominant between the Cascade River fault and the Twisp fault. Included in this metamorphic terrane are Permian rocks whose lithologies are similar to parts of the Cache Creek formation in British Columbia. Large Mesozoic and Tertiary batholiths intrude this terrane, again simulating cratonic conditions.

Between the Twisp and Pasayten faults (the Twisp and Eightmile fault of Eaton and Staatz, 1971) the rocks are volcanics and sediments of Upper Jurassic and Cretaceous age. These partial equivalents of the Gravina-Nutzotin belt (Berg, et al., 1972) represent the clearest tie between Cascade formations and rocks to the north, since Gravina-Nutzotin belt rocks appear intermittently from Alaska to the Columbia River Plateau.

East of the Pasayten fault is a large complex Mesozoic batholith. The fault itself appears to be a right lateral strike-slip structure with large but undetermined pre-Tertiary movement. The batholith is coeval with some Mesozoic rocks west of the Pasayten fault, suggesting a possible arc-trench-batholith relationship developed in the Upper Jurassic-Lower Cretaceous.

McBirney (1975) summarizes Cascade Cenozoic volcanism. In the porphyry copper province composition of volcanic rocks seems to reflect thickness of the lithosphere. Volcanism of the Cascade Range was concurrent with eposides elsewhere around the Circum-Pacific region and seems to have been independent of subduction rates or other local variables computed from a plate motion. Dating of Cascade porphyry copper deposits (from 6.2 to 49 m.y.) does not conflict with this conclusion.

Structure: Fig. 39 contains the same tectonic elements found in Fig. 34 for the Canadian Cordillera. The northwest trending Pinchi set is well represented in the Pasayten, Twisp, and Deception Creek faults, all of which may have had substantial right lateral movement. The most effective regional tectonic force acting intermittently through parts of the Upper Mesozoic and Lower Tertiary (as interpreted from this fault pattern and the dated igneous rocks cut by the pattern) was the northward movement of the Pacific plate relative to the North American plate. Movement between plates was accommodated by displacement on faults that make up the northwest-trending set.

Fold axes and thrust faults (not shown in Fig. 39) indicate that westward tectonic transport of the North American plate and intermittent eastward-dipping subduction below the plate also were effective in fracturing the North American plate margin. At this time it is impossible to tell which of these two phenomena was most important.

The spatial relationship between strike-slip faults and porphyry copper deposits suggests that the two are probably genetically related.

The strike-slip faults may have provided access to the surface and upper crust for material escaping from depth.

Fig. 39 also shows the southern boundary of Cascade tungsten and molybdenum occurrences (Culver and Broughton, 1945, and Purdy, 1954). This line coincides with the northern limit of the Pacific orogen (King, 1969). Both these lithophiles may be incorporated into Cascade porphyry systems, making porphyry deposits in this region unusual for the Cordilleran orogen. If the tungsten-molybdenum line is accepted as the southern limit of basement cratonic crustal conditions, conceivably porphyry deposits south of the line would differ distinctly from those to the north. Analysis of the statistically small number of deposits presented in Table 14 shows this may indeed be the case.

Porphyry deposits whose descriptions are published and are plotted in Fig. 39 form a northeast-trending zone anomalous to regional tectonic grain (northwest), the physiographic axis of the Cascades (north-south), and to the plate boundary (also northwest). Why these Tertiary deposits trend northeast remains to be explained.

Characteristics of Cascade Porphyry Copper Deposits

Table 14 summaries briefly the various characteristics found in Cascade porphyry copper deposits. Several striking features distinguish this province from porphyry copper provinces elsewhere in the Cordilleran orogen—i.e., the wide variety of both host and mineralized intrusive composition, the generally low sulfur potential of most deposits, the possibility that pyrrhotite may occur significantly in place of pyrite, and the common occurrence of tungsten as scheelite within the porphyry system. These features are examined in detail.

Petrography: Table 14 lists the composition of the intrusion closest to ore (modified from Grant, 1969). All are calc-alkalic and quartz-bearing. In many examples, particularly those north of the tungsten-molybdenum line, the pluton is an intrusive complex containing a dominant premineral quartz diorite phase. The smaller granodiorite and quartz monzonite phases close to mineralization are invariably younger and usually porphyritic. South of the tungsten-molybdenum line, both the pluton and the phase closest to ore (distinguished by textural features) tend to be quartz diorite.

In the quartz diorites many plutons have unusually calcic plagioclases (Grant, 1969). Hornblende is the most common mafic although biotite is also known. Most intrusive phases closely associated with mineralization are porphyritic. Quartz, hornblende, biotite, and zoned plagioclase of average andesine composition are the most commonly occurring phenocrysts, with orthoclase prominent in few deposits.

Where the plagioclase-orthoclase ratios approach one, orthoclase is more common in the groundmass of the porphyry. With plagioclase phenocrysts generally prominent, orthoclase tends to occur as a fine-grained interstitial mineral with quartz.

Alteration: The alteration sequence from the core of each deposit is shown in Table 14. All deposits have a potassic zone. Potassium metasomatism is ubiquitous but its effect varies from deposit to deposit. Some potassic zones are masses of velvety fine-grained secondary biotite with other silicates and sulfides comprising less than 20% of the rock (e.g., Mineral Creek). In others (e.g., Middle Fork) secondary orthoclase is dominant. Grant (1969) presents evidence of as much as 300% K_2O enrichment in the potassic zones in some deposits, but all examples used in his study occur north of the tungsten-molybdenum line. Examples of porphyry occurrences south or west of this line may have only minor secondary orthoclase with the biotite if it is present at all; secondary plagioclase may be the chief feldspar in the potassic zone. Chlorite also occurs erratically in the potassic zones as a hydrothermal mineral with other secondary silicates.

A typical quartz-sericite-pyrite phyllic zone occurs within most deposits. Occurrences with smaller pyrite halos may not have such an alteration assemblage, on the other hand, and the potassic zone may be ringed by a propylitic assemblage. However, where the phyllic zone does occur it is peripheral to the potassic. In all cases, sericite is a fine-grained pervasive silicate occurring with secondary quartz and pyrite.

A kaolin- or kaolin-illite-rich argillic zone appears peripheral to the phyllic zone in more than half the deposits. It varies in size and mineralogy but usually contains pervasive calcite as well as sulfide and clay. Its appearance could be interpreted as evidence of the young age and minimal erosion of deposits in this province.

A chlorite-rich propylitic zone surrounds the other alteration assemblages and also may contain epidote, pyrite, calcite, albite, and other typical propylitic zone minerals. Sulfides are not well developed, however.

Sulfur combining with iron in the potassic, phyllic, and argillic zones of Cascade deposits may develop pyrrhotite-rich halos rather than the typical pyritic halos more frequently seen in deposits elsewhere in the Cordilleran orogen. Pyrrhotite occurs rarely in smaller deposits elsewhere in the Cordilleran orogen (e.g., Battle Mountain). The frequency with which pyrrhotite porphyries are found in the Cascades sets this province apart. Both pyrite and pyrrhotite develop most strongly peripheral to the potassic zone rather than in it. The concentration of iron sulfide in the alteration zone adjacent to the potassic is common to all deposits regardless of the type of alteration (phyllic, argillic, or propylitic) that may be peripheral to the potassic zone. The sulfide occurs dominantly as a dissemination in the altered rock. Size of the iron sulfide halo is variable.

Tourmaline can be found as an alteration mineral in several deposits (Earl, Quartz Creek, Vesper).

Mineralization: The distinctive feature in the economic mineralogy of porphyry copper deposits of the Cascades is the possibility that tungsten (as scheelite) may occur with copper and molybdenum sulfides north of the tungsten-molybdenum line.

Scheelite occasionally is found in other porphyry copper deposits (e.g., Gaspe). As wolframite, tungsten has also been reported from a number of deposits in Mexico. The persistence of scheelite in numerous Cascade porphyry copper deposits makes this group of deposits unusual. Tungsten is known in 8 of the 12 porphyry copper examples north of the tungsten-molybdenum line shown in Table 14. All these deposits have visible scheelite occurring mostly as a dissemination if in the potassic or phyllic zone or as a fracture filling if found in the zones peripheral to the phyllic. Wolframite has been reported from a few deposits but is not ordinarily identified. Whether this is because it habitually occurs in fairly fine easily bypassed crystals, or whether it simply is not a common constituent has not been determined.

The possibility that scheelite may occur in all alteration zones suggests it is highly mobile. If zinc and lead occur outside the porphyry deposit in a metallogenically zoned district, scheelite has been reported to occur in the zinc-lead veins as well. Since these veins most commonly occur in the propylitic alteration zone the appearance of scheelite there is not an anomaly in the Cascades.

Although tungsten has not yet been reported in Cascade porphyry deposits south of the tungsten-molybdenum line, molybdenum has been found in deposits in sporadic subeconomic amounts. The continued (though diminished) presence of molybdenum coupled with an almost complete absence of tungsten south of this line coincides with an apparent shift in composition of the ore-associated intrusion. South of the tungsten-molybdenum line quartz diorite is the only plutonic type known to be associated with ore sulfides. The change in composition of the pluton concomitant with disappearance of scheelite can be explained by either a change in the composition or in the thickness of the sialic crust.

In many respects Cascade porphyry copper examples are similar to the Tertiary diapiric or stock type deposits of the Canadian Cordillera. They may be separated into Babine Lake or inside types containing mineralization largely within the plutonic rocks (e.g., Middle Fork) or alternatively into those deposits whose ore minerals occur mostly in a biotite-rich contact zone adjacent to a boss or stock, the Berg or outside type exemplified by North Fork in the Cascades. In the latter, minor mineralization may also appear within the pluton as well. Since most deposits appear to contain sulfide in the intrusive as well as the intruded rock no genetic distinction is possible for the two types.

Ore sulfides in a biotite-rich potassic zone usually are disseminated within this zone. Ore sulfides within the phyllic zone most commonly are fracture controlled. Very little copper or molybdenum sulfide occurs outside the phyllic zone and all deposits explored to date have only a small developed tonnage.

Mineral zoning with zinc-lead-silver-gold value's occurring in propylitically altered rocks adjacent to the copper zone typifies Cascade deposits. None of the zinc-lead deposits marginal to the porphyries has produced a significant tonnage of ore however.

Radiometric Age Relationships: Table 14 summaries radiometric dating for Cascade porphyry copper occurrences. In only a few cases have dates been determined for alteration silicates accompanying sulfide. The dates given are therefore largely for igneous rocks appearing to have a close time-space relationship to mineralization and mostly follow Armstrong, et al. (1976). No mineralized intrusions with Tertiary dates have been found east of the Pasayten fault and this fault may be considered the eastern boundary of this Tertiary province.

The Mazama deposit is located between a diorite porphyry and a quartz diorite porphyry in andesitic extrusives probably co-magmatic with the former. The diorite has been dated at 84 m.y. and the quartz diorite at 70 m.y. (Wolfhard, 1976). Mineralization appears to be associated with the quartz diorite stock. The Monument stock has a 49 m.y. K-Ar date. Porphyry mineralization cuts this pluton but may approach it in age. A less clear association exists for the Ross Lake-Davis deposit. The nearby Chilliwack batholith has a 30 m.y. K-Ar age. If mineralization coincided with the dated phase of the Chilliwack batholith, the Ross Lake-Davis deposit is also 30 m.y. old. Vesper (also

known as Sunrise or Bren-Mac) has been dated at 32 m.y. (Grant, 1976). Hydrothermal biotite with intergrown secondary quartz, plagioclase, and chalcopyrite has been found to have a 24 m.y. K-Ar date at McCoy Creek. This date appears to fix mineralization at that time. A clear case may be made for Glacier Peak as the Cloudy Pass pluton has been dated at 22 m.y. and the Glacier Peak ore deposit appears closely associated with it.

Quartz Creek, Middle Fork, and Mineral Creek deposits lie close to the Snoqualmie batholith, justifying speculative dating of these deposits at the same age as the pluton. With a K-Ar date of 18 m.y. the Snoqualmie batholith is one of the younger batholiths dated in the Cascades and the deposits near it may be assigned the same approximate dates.

The Earl deposit has a K-Ar date of 16.2 m.y. from hydrothermal biotite. A 21 m.y. date from magmatic biotite in the southern end of the Spirit Lake batholith about 10 miles south of the Earl is not considered to conflict with it.

North Fork near the Snoqualmie batholith has a K-Ar age of 9.9 m.y. from hydrothermal biotite. In both the North Fork and Earl datings the biotite was intergrown with chalcopyrite and therefore these dates are believed to represent ages of mineralization.

Hydrothermal sericite at Mesatchee provided a K-Ar date of 6.2 m.y.

The Cascade K-Ar dates 30, 24, 22, 18, 16, 9.9, and 6.2 m.y. agree with the geologic setting in each case. Apparent concordance between Cascade porphyry dates and Cascade volcanic rock suggest an episodic igneous history. Volcanic episodes at 0-2, 3-7, 9-11, and 14-18 m.y. have been recognized in central Oregon Cascade volcanic rocks and examples representative of each of these episodes exist in the Cascades porphyries. The 22 and 24 m.y. dates are nearly coincident with the Grotto (25.7 m.y.) and Monte Cristo (24.2 m.y.) stocks and early phases of the Tatoosh complex (Mattinson, 1973), suggesting another episode about 25 m.y. ago. The Earl (16.2 m.y.) date and dates for Snoqualmie and younger Tatoosh plutons are concordant with the 15-18 m.y. old Columbia volcanic episode. The 6.2 m.y. date for Mesatchee cannot be matched by other results from Washington but is similar to the 4-8 m.y. dates for the Laurel Hill stock. Mount Baker is a dormant volcano with a summit thermal-solfatara field that includes 78 000 sq m (9000 sq ft) of altered rock. Tephra associated with 1975 steam eruptions contains identifiable chalcopyrite as well as fairly common pyrite. The existence of chalcopyrite in this large currently developing alteration zone leads to speculation that a porphyry deposit is now forming within the Mount Baker thermal system.

No concensus has been reached on the question of whether or not subduction is active now or was a factor at the time the younger deposits formed in the Cascades.

Conclusions for Cascade Deposits

The Tertiary porphyry copper deposits west of the Pasayten fault have distinct alteration, mineralization, and age characteristics. They may include tungsten (scheelite) bearing phyllic and potassic zones. The deposits may contain pyrrhotite in place of or in addition to pyrite and may eventually be demonstrated to date as recently as the present. The deposits are typical stockwork or breccia structural types containing typical potassic zones and chalcopyrite and molybdenite are the most important ore minerals.

Known deposits appear to be located near or on one of several right lateral strike-slip faults. Movement of some of these faults may be speculated to be about 161-193 km (100-120 miles).

REFERENCES AND BIBLIOGRAPHY

Ager, C. A., McMillan, W. J., and Ulrych, T. J., 1972, "Gravity Magnetics and Geology of the Guichon Batholith," Bulletin No. 62, British Columbia Dept. of Mines, p. 112.

Armstrong, R. L., Harakal, J. E., and Hollister, V. F., 1976, "Late Cenozoic Porphyry Copper," *Transactions,* Canadian Institute of Mining and Metallurgy, Vol. 85.

Barr, D. A., 1966, "The Galore Creek Deposit," *Transactions,* Canadian Institute of Mining and Metallurgy, Vol. 69, p. 841.

Berg, H. C., Jones, D. L., and Richter, D. H., 1972, "Gravina-Nutzotin Belt," Professional Paper No. 800D, US Geological Survey, pp. D1-D24.

Burchfiel, B. C., and Davis, G. A., 1973, "Possible Igneous Analog at Salt Dome Techtonics, Clark Mountain, Southern California," Bulletin No. 9, American Association of Petroleum Geologists, p. 933.

Carr, J. M., 1976, "Supergene Alteration at Afton," *Abstracts,* Geological Association of Canada, Vol. 1976, Cordilleran Section.

Carson, D. J. T., and Jambor, J. T., 1974, "Porphyry Copper Deposits, Babine Lake Area, British Columbia," *Transactions,* Canadian Institute of Mining and Metallurgy, Vol. 76, p. 1.

Carter, N. C., and Kirkham, R. V., 1969, "Geological Compilation Map of the Smithers, Hazelton, and Terrace Areas," Map No. 69-1, British Columbia Dept. of Mines.

Carter, N. C., 1974, "Geochronology of Porphyry Deposits of Central British Columbia," Ph.D. thesis, University of British Columbia, Vancouver, B. C., Canada.

Clark, K. F., 1972, "Stockwork Molybdenum Deposits in the Western Cordillera," *Economic Geology,* Vol. 67, pp. 731-758.

Culver, H. E., and Broughton, W. A., 1945, "Tungsten Resources of Washington," Bulletin No. 34, Washington Div. of Mines and Geology.

Drummond, A. D., Tennant, S. J., and Young, R. J., 1972, "The Interrelationship of Regional Metamorphism, Hydrothermal Alteration and Mineralization at Gibraltar," *Transactions,* Canadian Institute of Mining and Metallurgy, Vol. 74, p. 218.

Eaton, G. P., and Staatz, M. H., 1971, "Mineral Resources of the Pasayten Wilderness Area, Washington," Bulletin No. 1325, US Geological Survey.

Field, C. W., Jones, M. B., and Bruce, W. R., 1974, "Porphyry Copper-Molybdenum Deposits of the Pacific Northwest," *Trans., SME-AIME,* Vol. 255, p. 9.

Fox, P. E., 1975, "Alkaline Rocks and Related Deposits," *Abstracts,* Intrusive Rocks of Mineral Deposits, Geological Association of Canada, Cordilleran Sec.

Gaultieri, J. L., et al., 1973, "Mineral Resources of the Alpine Lakes Study Area, Washington," Open File Report, US Geological Survey.

Grant, A. R., 1969, "Chemical and Physical Controls for Base Metal Deposition in the Cascade Range," Bulletin No. 58, Washington Div. of Mines and Geology.

Grant, A. R., 1976, "Mineral Resource Analysis Study on US Forest Service Land, Washington," Contract No. 004724N, US Forest Service, Washington, DC.

Gussow, W. D., 1962, "Energy Source of Intrusive Masses," Transactions, Royal Society of Canada, Vol. 61, Series 3, pp. 1-19.

Hill, D. P., 1975, "Crustal and Upper Mantle Structure," Abstracts, Geological Society of America, Vol. 7, p. 1116.

Hollister, V. F., 1974, "Evolution of the Porphyry Copper Province of the Northern Cordilleran Orogen," Proceedings, 1st Circum-Pacific Energy and Mineral Resources Conference, American Association of Petroleum Geologists, 1976.

Hollister, V. F., Anzalone, S. A., and Richter, D. H,. 1975, "Porphyry Copper Belts of Southern Alaska and Contiguous Yukon," Bulletin No. 4, Canadian Institute of Mining and Metallurgy, p. 51.

Hollister, V. F., et al., 1975, "Structural Evolution of Porphyry Mineralization at Highland Valley," Canadian Journal of Earth Sciences, Vol. 12, No. 5, p. 579.

Huntting, M. T., 1956, "Metallic Minerals, Part 2," Inventory of Washington Minerals, Bulletin No. 37, Washington Div. of Mines and Geology, p. 428.

Hylands, J. J., 1972, "Porphyry Copper Deposits of the Guichon Batholith," 24th International Geological Congress, Sec. 4, p. 241.

Jones, D. L., Irwin, W. P., and Ovenshine, A. T., 1972, "Southeastern Alaska—A Displaced Continental Fragment," Professional Paper No. 800B, US Geological Survey, p. B211.

King, P. B., 1969, "The Tectonic Map of North America," Map No.   , US Geological Survey.

Larson, R. L., and Pitman, C. G., 1972, "Late Mesozoic and Cenozoic Plutonic Events in Circum-Pacific North America," Bulletin, Geological Society of America, Vol. 84, p. 3773.

Lowell, J. D., and Guilbert, J. M., 1970, "Lateral and Vertical Alteration-Mineralization Zoning in Porphyry Copper Deposits," Economic Geology, Vol. 65, pp. 373-408.

Mattinson, J. M., 1972, "Ages of Zircons from the Northern Cascades, Washington," Bulletin, Geological Society of America, Vol. 83, p. 3769.

McBirney, A. R., 1975, "Volcanic Evolution of the Cascade Range," Abstracts, Geological Society of America, Vol. 7, p. 1192.

McCarthy, T. S., and Jacobsen, J. B. E., 1976, "The Mineralizing Fluids at Artonvilla," Economic Geology, Vol. 71, p. 131.

McKee, B., 1972, Cascadia, McGraw-Hill, Inc., New York, NY, p. 390.

Misch, P. H., 1966, "Geology of the Northern Cascades," A Symposium on the Tectonic History and Mineral Deposits of the Western Cordillera, Special Volume 8, Canadian Institute of Mining and Metallurgy, pp. 101-148.

Monger, J. W. H., and Ross, C. A., 1971, "Distribution of Fusulinaceans in the Western Canadian Cordillera," Canadian Journal of Earth Sciences, Vol. 8, No. 2, p. 259.

Monger, J. W. H., Souther, J. G., and Gabrielse, H., 1972, "Evolution of the Canadian Cordillera, A Plate Tectonic Model," American Journal of Science, Vol. 272, pp. 577-602.

Monger, J. W. H., 1975, "Correlation of Eugeosynclinal Tectono-Stratigraphic Belts in the North American Cordillera," Geoscience Canada, Vol. 2, No. 1, p. 4.

Ney, C. S., and Brown, A. S., 1972, "Copper and Molybdenum Deposits of the Western Cordillera," Excursion No. A09, 24th International Geological Congress.

Northcote, K. E., 1969, "Geology and Geochronology of the Guichon Creek Batholith," Bulletin No. 56, British Columbia Dept. of Mines, p. 219.

Paterson, I. A., and Harakal, J. E., 1974, "Potassium-Argon Dating of Blueschists from Pinchi Lake, British Columbia," Canadian Journal of Earth Sciences, Vol. 11, p. 1011.

Patton, T. C., Grant, A. R., and Cheney, E. S., 1973, "Hydrothermal Alteration at the Middle Fork Copper Prospect, Central Cascades, Washington," Economic Geology, Vol. 68, p. 816.

Porphyry Deposits of the Canadian Cordillera, A. Sutherland Brown, ed., Special Volume 15, Canadian Institute of Mining and Metallurgy, 1976.

Preto, V., 1975, "The Nicola Group: Mesozoic Volcanism Related to Rifting," Abstracts, Geological Society of America, Vol. 7, No. 6, p. 820.

Purdy, C. P., 1954, "Molybdenum Occurrences of Washington," Report Inventory No. 18, Washington Div. of Mines and Geology, p. 118.

Ross, C. A., 1967, "Development of Fusulinid Faunal Regimes," Journal of Paleontology, Vol. 41, No. 6, p. 1341.

Seraphim, R. H., and Hollister, V. F., 1975, "Structural Setting of Porphyry Deposits of the Canadian Cordillera," TP-2C, Canadian Institute of Mining and Metallurgy Annual Meeting.

Simson, G. C., Van Noy, R. M., and Zilka, N. T., 1974, "Mineral Resources of the Cougar Lakes Study Area," Open File Report No. 74-243, US Geological Survey.

Soregaroli, A. E., 1971, "Geology of the Brenda Copper-Molybdenum Deposit," TP-18, Canadian Institute of Mining and Metallurgy 73rd Annual Meeting, p. 17.

Soregaroli, A. E., 1975, "Geology of Porphyry Copper and Molybdenum Deposits," Paper No. 75-1, Geological Survey of Canada, p. 244.

Stacey, R. A., 1974, "Deep Structure of Porphyry Ore Deposits in the Canadian Cordillera," Contribution No. 504, Dept. of Energy, Mines, and Resources, Earth Physics Branch, Ottawa, Ont., Canada.

Sutherland Brown, A., et al., 1971, "Metallogeny of the Canadian Cordillera," *Transactions,* Canadian Institute of Mining and Metallurgy, Vol. 74, pp. 1-25.

Sutherland Brown, A., 1972, "Morphology and Classification of Porphyry Deposits of the Canadian Cordillera," Western Inter-University Geological Conference, University of British Columbia, Vancouver, B.C., Canada.

Tabor, R. W., 1963, "Large Quartz Diorite Dike and Explosion Breccia, Northern Cascade Mountain, Washington," *Bulletin,* Geological Society of America, Vol. 74, p. 1203.

Wanless, R. K., 1969, "Isotopic Age Map of Canada," Map No. 1256A, Geological Survey of Canada.

Wheeler, J. O., et al., 1972, "The Cordilleran Structural Province," *Variations in Tectonic Styles in Canada,* Special Paper No. 11, Geological Association of Canada.

White, W. H., 1959, "Cordilleran Tectonics in British Columbia," *Bulletin,* American Association of Petroleum Geologists, Vol. 43, pp. 60-100.

Wolfhard, M., and Ney, C., 1974, "Metallization in the Canadian Cordillera," Special Paper No. 14, Geological Association of Canada, in press.

Wolfhard, M., 1976, Private Communication.

5

Porphyry Copper Deposits of the Southern Cordilleran Orogen

CONTENTS

INTRODUCTION

The geology of porphyry copper deposits in the southern Cordilleran orogen is the subject of this chapter. The area considered extends northwest from southern Mexico to the Columbia volcanic plateau, the Cascade volcanic province, and the Idaho batholith. Transversely it is a belt some 800 km (500 miles) in width bounded on the west by the Sierra Nevada range. Included within this area are many large segments considered unfavorable for hosting porphyry copper deposits, which are mentioned in the text. The region and its porphyry copper deposits have probably been more intensely studied and reported in the literature than any other area in the world. This summary considers regional geologic controls that governed formation of porphyry copper deposits within this province. Geologic factors that influenced the mineralogy and metallogeny of the deposits are also examined.

Discussions with H. R. Cornwall have greatly helped in writing this section.

GEOLOGIC SETTING

The geologic setting has been summarized by Anderson (1966), Burchfiel and Davis (1972), Rogers, et al. (1974), and Lowell (1974). No porphyry copper deposits in this area are known to be pre-Triassic, so Mesozoic and Tertiary events will be stressed. Figs. 40 and 41 show basic data believed most pertinent to the topic of porphyry copper deposits.

Pre-Middle Jurassic: Figs. 40 and 41 contain the principal structural and crustal elements found in the Cordilleran orogen after Anderson (1966), Schmitt (1966), Coney (1972), Rogers, et al. (1974), and Lowell (1974, 1976), together with location of some of the more important porphyry copper deposits. The western limit of the Precambrian craton is shown in Fig. 40 (Rogers, et al., 1974). East of that limit the crustal environment through which the porphyry copper plutons ascended includes the Precambrian basement as well as varying thicknesses of Paleozoic shelf and platform accumulations.

The northwestern and western portion of the southern Cordilleran orogen west of the Rogers, et al. (1974) limit of Precambrian basement has what Churkin (1974) calls a thin continental crust. Brooks (1976) and Churkin (1974) describe oceanic volcanic

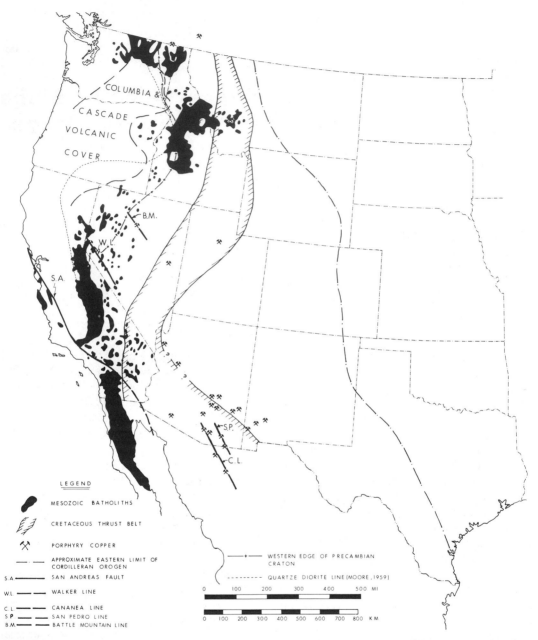

Fig. 40. Batholiths, thrust belts, and porphyry copper belts of the southern Cordilleran orogen. Porphyry copper deposits of the southern Cordilleran orogen tend to occur in distinct belts east of the large Mesozoic batholiths. Moore's (1959) line and the western limit of the Precambrian craton (Rogers, et al., 1974) lie west of all large Lowell and Guilbert (1970) model porphyry copper deposits now known.

DEPOSITS

1. SANTA RITA
2. TYRONE
3. MORENCI
4. BISBEE
5. CANANEA
6. MIAMI - INSPIRATION
7. RAY
8. SAN MANUEL
9. SILVERBELL
10. PIMA - MISSION
11. ESPERANZA
12. BAGDAD
13. AJO
14. PILARES
15. CARIDAD
16. ITHICA PEAK
17. YERINGTON
18. BATTLE MOUNTAIN
19. BINGHAM
20. ELY
21. BUTTE
22. KERWIN
23. LIGHTS CREEK
24. COPPER BASIN
25. EL ARCO

LEGEND

⚒ - Major Porphyry Copper
     (and those mentioned in Text)

o - Minor Porphyry Copper

⚒ - 200 Milligals or less.

0   100   200   300   400   500 MI

0 100 200 300 400 500 600 700 800 KM

Fig. 41. Porphyry copper deposits and gravity. Crustal thickness implied by gravity data with distribution of porphyry copper prospects superimposed suggests that the absence of porphyry copper deposits from areas of thick crust is a valid exploration parameter (*modified after Lyons, 1950*).

assemblages there that include ultrabasic rocks as significant components of the crust. Burchfiel and Davis (1972) note that the upper Paleozoic ophiolite sequence is overlain from Canada to Mexico by a Lower Mesozoic, largely marine volcanic sedimentary complex, which developed during and after a Permo-Triassic deformation that established a northwest tectonic grain in California, Arizona, and Mexico. These rocks covered or cut all previous structural trends.

Periods of significant strike-slip faulting are inferred in the development of the pre-Middle Jurassic by Silver and Anderson (1974), Jones (1972), Burchfiel and Davis (1972), and Coney (1975). No concensus has been established on type (left lateral or right lateral) or amount of movement. The metallogeny of Upper Mesozoic and Tertiary porphyry copper deposits seems to be influenced by pre-Middle Jurassic crustal evolution. Significantly, most porphyry copper deposits lie east of the Rogers, et al. (1974) western edge of the Precambrian craton and all porphyry copper deposits that contain recoverable byproduct molybdenum lie east of their line. Deposits occurring west of the Rogers, et al. (1974) western limit of the Precambrian have an anomalously high gold:copper ratio. Some copper-gold deposits also formed east of the western limit of the Precambrian in special tectonic settings. The composition of the crust established by mid-Jurassic influenced the kind of metals found in the mineralized plutons that penetrated it during the Mesozoic and Tertiary.

Local orogenies recorded in the Paleozoic in this area were accompanied by large-scale thrust faulting (Burchfiel and Davis, 1972; Monger, et al., 1972). No porphyry copper deposits have been found associated with pre-Middle Jurassic thrusts and therefore these Paleozoic structures are omitted from Figs. 40 and 41.

Bisbee may have formed during this period, with a K-Ar date of 163 m.y. (Lowell, 1974), although Bryant (1968) suggests a 130 m.y. K-Ar date. A distensional setting is indicated for Bisbee because a marked grabenlike subsidence trough developed in this area during the period that includes intrusion and mineralization.

Middle Jurassic to Late Upper Cretaceous: Atwater (1970), Coney (1972), and Burchfiel and Davis (1972) imply that at least two arc-trench systems that existed in the western Cordilleran orogen in the Middle and Upper Jurassic and Lower Cretaceous may have been part of one continuous system. Large batholithic masses developed with these and Armstrong and Suppe (1973) propose that Franciscan trench deposition spanned the interval 150 to 70 m.y.

Larson and Pitman (1973) have shown that batholithic intrusion occurred in a magnetic quiet period from 110 to 85 m.y., which they assumed to have been a period of accelerated sea floor spreading and rapid subduction. Folding along northwest axes also occurred during this period. The Cretaceous thrust belt (Fig. 40, Anderson, 1966) developed during this interval. Rapid subduction at a plate-consuming margin accompanied by large volumes of intrusive and extrusive material apparently did not provide a favorable environment for porphyry copper development, however, because few porphyry copper deposits are known from this island arc-trench melange-pluton assemblage. Most deposits dated close to or within this interval occur well within the cratonic block and away from the arc-trench environment. Yerington and the nearby Ann-Mason (Mickey Pass) ore bodies probably have K-Ar 111 m.y. dates (Schilling, 1972). These ore bodies are elongate within a zone of faulting that has many appearances of a major strike-slip fault. Internal structure within the Yerington district is compatible with right lateral movement on a major northwest trending structure nearby at the time of mineralization. Strike-slip faulting may be inferred to have existed in the Cretaceous at time of mineralization at Yerington, about 111 m.y. ago. It seems reasonable to speculate that such features as the Walker line (in the sense of Lowell, 1974) were active during mineralization at Yerington (Fig. 40). Rogers, et al. (1974), show a right lateral displacement (Fig. 40) for the Precambrian craton approximately where the Walker line exists, though age of this displacement is not specified.

Highly altered porphyry from Ely also has a 111 m.y. K-Ar date, preceding the Larson and Pitman (1973) magnetic quiet zone with its simultaneous rapid subduction, folding, and more intense batholithic activity. Alternative and widely accepted dates for both Ely and Yerington are quoted at 109 m.y. Few porphyry copper deposits, however, are clearly dated radiometrically within the Late Jurassic-Lower Cretaceous magnetic quiet interval and a negative correlation does appear to emerge between batholith formation and rapid subduction on one hand and porphyry copper formation on the other.

El Arco, in Baja California occurs with an intrusive-extrusive igneous suite with K-Ar ages spanning the 117-107 m.y. range. Mineralization appears associated with an intrusion dated at 107 m.y. (Barthelmy, 1974). This diorite model porphyry copper is more fully described later in this chapter. Mineralization is stated to be post-tectonic deformation despite its anomalous date.

Late Upper Cretaceous to Upper Miocene: Burchfiel and Davis (1972) show a marked change in tectonic style to have occurred about 80 m.y. ago. Larson and Pitman (1973) suggest the change to have taken place at about 85 m.y. based on magnetic data from ocean basins. Prior to this date, crustal foreshortening (e.g., folding and large-scale thrust faulting) and batholithic intrusion were dominant in the Cordilleran orogen. Right lateral strike-slip faulting on a major regional scale became important as compressive stresses weakened. Most porphyry copper deposits date from that abrupt change.

Whereas Burchfiel and Davis (1972) mention the inception of north-northwest right lateral faulting with the relaxation of compression, Rehrig and Heidrik (1976) interpret the systematic fracture pattern found in porphyry copper deposits to "indicate the presence of widespread and temporally consistent crustal stress fields" during mineralization. The dilational nature of these fractures filled with dike and stockwork vein swarms imply a distensional tectonic setting for the 75-50 m.y. period when most Arizona porphyry mineralization formed. Rehrig and Heidrik (1976) propose a regional crustal lengthening oriented north-northwest by south-southeast for this time interval. Crustal distension and porphyry copper mineralization continued, with the youngest deposit dated at 20 m.y.

Extensive Tertiary volcanic and tectonic overprinting has obscured earlier demonstrably different tectonic styles in the Basin and Range physiographic province. Basin-Range faulting, if defined to include all normal faults within this province, began not later than early Tertiary time (Lowell, 1974). Differential uplift and subsidence became more prominent about 30 m.y. ago and extended sharply through all the Basin and Range province by about 17 m.y. (Stewart, 1971). Lowell (1974) notes the presence of mid-Tertiary thrust faults that developed during this period of strike-slip tectonics within the porphyry copper province, but he suggests gravity gliding as a more important cause than regional compression. Drewes (1976) concurs that gravity glide faults developed in the mid-Tertiary. The regional tectonic frame-work for this period described by Burchfiel and Davis (1972), Lowell (1974), and Drewes (1976) is compatible with the Rehrig and Heidrik (1976) proposal for crustal distension.

Although porphyry copper deposits formed in the period of strike-slip faulting, few deposits have been identified as occurring directly on regional strike-slip megastructures. Silver Bell (Richard and Courtright, 1961) may be an exception, as may be some deposits on the Walker line, the San Pedro line (Lowell, 1974), and Bingham (Moore and Nash, 1974).

Plutonism in this period was largely restricted to smaller stocks, many of which were associated with volcanic piles. Many of these intrusions are structurally controlled in some manner, although they appear to have a more or less random distribution when plotted on a map. Spatial distribution of plutonism associated with porphyry copper deposits alone is less random, however (Fig. 41).

Dating of all plutons shows fairly erratic time distribution through the Tertiary. On the other hand, if the Cascade dates are excepted as relating to a late Tertiary porphyry copper province the remaining dates for intrusions show a progressively younger trend to the east. Porphyry copper dates alone, however, do not coincide with this trend (Noble, 1974), as deposits occur independently within the orogen.

Fig. 41 shows the location of most porphyry copper type deposits regardless of size. Deposits in this figure include those containing at least 18 million mt (20 million tons) of 0.1% Cu and possessing porphyry type alteration (Lowell, 1974). Most of these formed from the Upper Cretaceous to the upper Miocene. Only the most widely investigated and reported deposits, however, are shown in Fig. 40 and Table 15.

Post-Upper Miocene: There is little dissension to the widely held belief that Late Tertiary tectonics of the Great Basin were characterized by absence of crustal compression (Scholz, et al., 1971). Horst and graben structures of the Basin and Range province are based on deep-seated extension (Stewart, 1971) and the broad scale development of horst-graben structure was accompanied by a change in the composition of igneous activity to bimodal volcanism (Noble, 1972) with basaltic lavas becoming most common in the last 10 m.y. Accentuation of Basin and Range development about 17 m.y. ago (Burchfiel and Davis, 1972) seems to have marked the end of porphyry copper mineralization (Lowell, 1974).

Activity on the Walker line was rejuvenated in the post-17 m.y. period (Lowell, 1974) but porphyry copper type mineralization found close to this line from Ithica Peak (Mineral Park), AZ, to Yerington, NV, is clearly pre-upper Miocene.

Table 15. Principal Porphyry Copper Deposits of the Southern Cordilleran Orogen

	Name	Age, m.y.	Intrusion	Alteration Zones from Core	Structure
1.	Santa Rita	63	Quartz Mon	Pot-Phy-Arg-Prop	Stockwork
2.	Tyrone	56	Quartz Mon	Pot-Phy-Arg-Prop	Stockwork
3.	Morenci	55	Quartz Mon	Phy-Arg-Prop	Stockwork
4.	Bisbee	163	Grdr	Phy-Arg-Prop	Breccia
5.	Cananea	59	Quartz Mon	Phy-Arg-Prop	Breccia
6.	Miami-Inspiration	60	Quartz Mon	Pot-Phy-Arg-Prop	Stockwork
7.	Ray	63	Quartz Mon	Pot-Phy-Arg-Prop	Stockwork
8.	San Manuel	67	Quartz Mon	Pot-Phy-Arg-Prop	Stockwork
9.	Silver Bell	63	Quartz Mon	Pot-Phy-Arg-Prop	Stockwork
10.	Pima-Mission	60	—	Skarn	Stockwork
11.	Esperanza	62	Quartz Mon	Pot-Phy-Arg-Prop	Stockwork
12.	Bagdad	71	Quartz Mon	Pot-Phy-Prop	Stockwork
13.	Ajo	63	Quartz Mon	Pot-Phy-Prop	Stockwork
14.	Pilares	53	Quartz Mon	Phy-Arg-Prop	Breccia
15.	Caridad	53	Quartz Mon	Phy-Arg-Prop	Breccia
16.	Ithica Peak	72	Quartz Mon	Pot-Phy-Arg-Prop	Stockwork
17.	Yerington	111	Quartz Mon	Pot-Phy-Prop	Stockwork
18.	Battle Mountain	39	Quartz Mon	Pot-Phy-Arg-Prop	Stockwork
19.	Bingham	37	Granite Por	Pot-Phy-Arg-Prop	Stockwork
20.	Ely	111	Quartz Mon	Pot-Phy-Prop	Stockwork
21.	Butte	69	Quartz Mon	Pot-Phy-Arg-Prop	Stockwork
22.	Kerwin	—	Quartz Mon	Pot-Phy-Arg-Prop	Stockwork
23.	Lights Creek	—	Quartz Dio	Pot-Prop	Stockwork
24.	Copper Basin	64	Quartz Mon	Pot-Phy-Arg-Prop	Stockwork
25.	El Arco	107	Mon-Syenite	Pot-Prop	Stockwork

Grd: Granodiorite	Dio: Diorite	Phy: Phyllic	Prop: Propylitic
Mon: Monzonite	Pot: Potassic	Arg: Argillic	

CHARACTERISTICS OF PORPHYRY COPPER DEPOSITS OF THE SOUTHERN CORDILLERAN OROGEN

Petrography, alteration, mineralization, and structural characteristics of a representative group of porphyry copper deposits are shown in Table 15. Deposits selected for this table contain features believed representative of other large deposits in this province. None of the smaller deposits are included in Table 15; therefore the table is not entirely representative of characteristics of smaller deposits. Generally, the smaller the deposit the less complete the development of alteration zoning. Smaller deposits also tend to contain a lower grade hypogene copper sulfide as well as a less extensive pyrite halo. Because the distinctions consist of omission in the smaller deposits and these vary from one occurrence to another, they cannot be easily tabulated.

The following sections on petrography, alteration, mineralization, and structure provide details for the summary outline given in Table 15.

Petrography

Stringham (1966) points out that a number of intrusive phases exist in the plutonic complexes of most major porphyry copper districts. Commonly granitoid rocks formed first and porphyritic rocks last. A progression also appears to exist from the most basic phase (usually quartz diorite) to the mineralized phase, a progression that favors hypothetical magmatic differentiation and ultimate derivation of copper from differentiating magma. Composition of the phase that is volumetrically most important may range from quartz diorite to granite, but a composition close to the granodiorite-quartz monzonite line occurs most commonly. Fig. 42, modified from Creasey (1966), is a ternary diagram summarizing composition of some unmineralized portions of larger and better known complexes. Compositions shown are of fresh rock and are either modal or normative, depending on which is available in the literature. Such diagrams commonly do not include both normative and modal analyses,

but the very broad purposes served by this figure permit such generalization. Figs. 36a and 36b (pages 103, 104) provide a comparison with ternary diagrams from other areas of the Cordilleran orogen. Fresh rock analyses alone are used; therefore they may not be completely representative of the mineralized phase. The composition indicated in Fig. 42 may differ from that shown in Table 15.

Ore occurring in a complex accompanies alteration of the rocks that host it as well as the phase of the pluton close to it in time. The majority of magmas considered spatially and temporally close to porphyry copper mineralization are porphyritic quartz phenocryst-bearing intrusions probably of quartz monzonite composition. Thus the composition indicated in Fig. 42 may not necessarily reflect the composition of the phase temporally or spatially closest to ore. The consistent spatial relationship between a younger porphyry phase and ore indicates a common source for both and implies a close time relationship.

Typically the plutons represented in Fig. 42 contain quartz, hornblende, biotite, plagioclase, and orthoclase. Microcline is very rare. The plagioclase is nearly always zoned with phenocrysts commonly averaging oligoclase-andesine in composition. Orthoclase may occur in both the groundmass and in phenocrysts although not all porphyries carry the orthoclase as phenocrysts. Biotite and hornblende both tend to occur as phenocrysts and may have a ratio of one or vary to the point where either biotite or hornblende is almost absent. Pyrite, chalcopyrite, and molybdenite each may rarely occur as accessory minerals along with the more common rutile, apatite, sphene, and magnetite.

Whereas phenocrysts are euhedral to subhedral, groundmass minerals are usually anhedral. Groundmass minerals are also randomly oriented and commonly equidimensional.

Many porphyry copper districts also contain diabase as part of the intrusive complex and in some cases this is well mineralized.

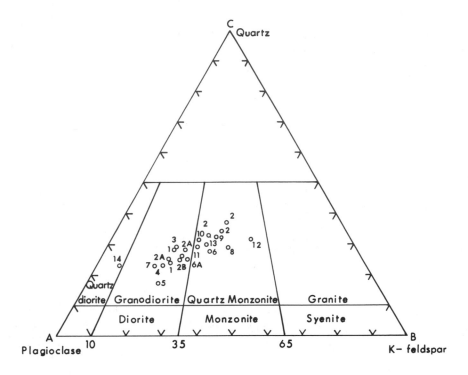

1. Cornelia Quartz Monzonite, Ajo, Arizona.
2. Lost Gulch Quartz Monzonite, Globe—Miami, Arizona.
2A. Schultze Granite, Globe—Miami, Arizona.
2B. Granite porphyry, Globe—Miami, Arizona.
3. Granite Mountain Porphyry, Ray, Arizona.
4. Granodiorite porphyry, San Manuel, Arizona.
5. Quartz monzonite, Bagdad, Arizona.
6. Quartz monzonite, Pima, Arizona.
6A. Granodiorite, Pima, Arizona.
7. Quartz monzonite porphyry, Morenci, Arizona.
8. Quartz monzonite, Battle Mountain, Nevada.
9. Quartz monzonite, Ithica Peak, Arizona.
10. Quartz monzonite, Butte, Montana.
11. Quartz monzonite, Kerwin, Wyoming.
12. Granite porphyry, Bingham Canyon, Utah.
13. Quartz monzonite, Santa Rita, New Mexico.
14. Quartz Diorite, Lights Creek, California.

Fig. 42. Petrography of southern Cordilleran deposits. Modal and normative analyses of fresh igneous phases from porphyry copper centers in the southern Cordilleran orogen show a strong grouping near the granodiorite-quartz monzonite boundary. Although they are not commonly merged, both modal and normative analyses are shown here since the diagram is used for general purposes only (*modified after Creasey, 1966*).

James (1971) mentions the tendency of diabase to be selectively mineralized in several districts, regardless of the age of the rock.

At El Arco, however, the ore is associated with andesite intruded by syenite (Barthelmy, 1974), which is the youngest phase of a diorite-syenite zoned intrusion. Quartz is absent in megascopic examination of the host rock and in this respect El Arco is unusual for the southern Cordilleran orogen. In a broader aspect, the syenite is part of a granodiorite-dominant plutonic complex that intrudes its own marine andesitic volcanic pile (Allisistos Formation). Alteration minerals are zonally aranged with a core potassic zone surrounded by a propylitic zone. Copper occurs in both. The intrusion is therefore co-magmatic with the andesitic rocks. Gold occurs significantly in this deposit as evidenced by old gold-placer workings over and near the copper-bearing outcrop.

Moorbath, et al. (1967) found initial strontium 87-strontium 86 ratios in the 0.706 to 0.708 range for Santa Rita and Ray and suggested that this indicates a mantle derivation of these elements for those plutons. Lowell (1974) quotes other such ratios, expanding the number of porphyry copper deposits appearing to have a subcrustal origin. A mantle source for porphyry magma would be compatible with the general geologic setting (Lowell, 1974) and may explain some independence between magma composition and upper crustal rock. Hedge (1974) notes that crustal contamination is detectable in porphyry copper magmas using strontium ratios determined for Colorado magmas. The mixed mantle-crustal source for the magma is confusing to those seeking the source of the metals and the magma.

Lead isotope ratios from the southern Cordilleran porphyry deposits and associated genetically related rocks are commonly found to be rather similar and distinctly less radiogenic than the crust invaded. The more enriched in radiogenic lead isotopes a rock is, the greater the contribution of upper crustal sources. Hence such intrusions are likely to be considered as mantle derived with some crustal assimilation from lead isotopic data. The distinctive isotopic characteristics of lead in subducted material makes it an unlikely source of lead in most porphyry copper deposits (Zartman, 1974).

Alteration

Hypogene alteration studies have been published for each of the larger deposits. Individual descriptions have been compiled into general summaries by Rose (1970), Lowell and Guilbert (1970), and Meyer and Hemley (1967). These compilations are schematically shown in Fig. 43, which is an ideal characterization of hypogene alteration zones in host rocks containing excess silica whose composition approximates granodiorite. Rose (1970) clearly shows that at any particular deposit one or more zones shown in Fig. 43 may be missing. Hypogene alteration is a metasomatic process not a metamorphism, and whole rock ratios of potassium to sodium, calcium, and magnesium are changed in the process. Supergene alteration (Rose, 1970) may mask or emphasize some key minerals in each hypogene zone so that all the zones present at depth may not be visible at the surface.

Hypogene Alteration: Fig. 43 depicts alteration zoning as it may be developed ideally from a potassic core through phyllic, argillic, and propylitic zones in a large porphyry copper deposit of the southern Cordilleran orogen (Lowell and Guilbert, 1970). In detail most deposits exhibit marked skewing of the zones in some direction (Rose, 1970).

Pyrite is a highly variable constituent in all zones. It may be present ranging from 1 to 3 wt % in the potassic zone, from 5 to 20 wt % in the phyllic, from 1 to 4 wt % in the argillic, and from 0 to 2 wt % in the propylitic. From its peak in the phyllic zone the pyrite content gradually weakens inward toward the potassic core but diminishes abruptly outward toward the propylitic zone (Fig. 43). The pyrite halo usually is developed around an intrusive center or some identifiable heat source. In most southern Cordilleran porphyry copper deposits the controlling feature is a definable stock. The most intense pyritization tends to rim the

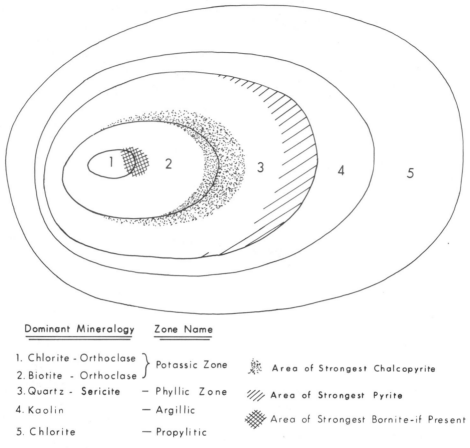

Dominant Mineralogy	Zone Name	
1. Chlorite - Orthoclase ⎫	Potassic Zone	![symbol] Area of Strongest Chalcopyrite
2. Biotite - Orthoclase ⎭		
3. Quartz - Sericite	— Phyllic Zone	![symbol] Area of Strongest Pyrite
4. Kaolin	— Argillic	![symbol] Area of Strongest Bornite-if Present
5. Chlorite	— Propylitic	

Fig. 43. Hydrothermal alteration zones and hypogene mineralization. All hypogene alteration zones are shown on this generalization of the Lowell and Guilbert (1970) model although in fact few deposits have every zone well developed. Deeply eroded deposits in particular may not have an exposed argillic zone and very deeply eroded deposits may not contain a phyllic zone. Phyllic zones may also be absent from dry deposits. Areas of pyrite and chalcopyrite concentrations are shown where they exist for many, but not all, deposits in this model.

zone of strongest hypogene copper mineralization.

Iron not consumed by sulfur in the alteration sequence may combine to form silicates (biotite, chlorite, or amphibole in the potassic zone or chlorite or epidote in the propylitic) or occur as magnetite in either the potassic or propylitic or both. Pyrrhotite is not common although it may accompany pyrite (e.g., Battle Mountain, NV).

The mode of occurrence of pyrite also changes from zone to zone, tending to appear largely in fractures in the propylitic zone but mostly as a dissemination in the potassic, phyllic, and argillic zones.

In areas where porphyry type deposits are common, pyrite halos and zones of pervasive pyritization may occur in a linear arrangement. Fig. 45 shows one such example that includes Caridad, Pilares, and Cananea. Areas of pyritization occur between these deposits, suggesting some subtle structural and mineralization continuity between deposits.

Apart from pervasive pyritization, metasomatically derived silicate mineral assemblages have been grouped to define and classify alteration zones (Lowell and Guilbert, 1970; Rose, 1970; and Creasey, 1959). The core zone is defined as the potassic (Creasey, 1959) and may be characterized either by secondary orthoclase and chlorite (e.g., Kalamazoo) or biotite and orthoclase (Rose, 1970). A secondary amphibole-rich zone (e.g., Bingham) may occur central to a biotite or biotite-chlorite zone. The orthoclase-chlorite subzone is missing from the surface exposure of most deposits. Montmorillonite, magnetite, quartz, sericite, sodium plagioclase, pyrite, and kaolinite have also been reported in varying but subordinate amounts. Paragenetic studies show that a typical sequential appearance of secondary ferromagnesium silicates is biotite, chlorite, amphibole (Moore and Nash, 1974) if the potassic zone contains more than one of these minerals.

Sheppard, et al. (1971) and Taylor (1974) have found through stable isotope techniques that hydrothermal biotite from the potassic zones at Santa Rita, Ely, and Bingham contain D/H (deuterium/hydrogen) ratios indicating dominance of magmatic water at time of formation. The O^{18} shift for quartz-orthoclase-biotite assemblages in these deposits also supports the magmatic affinity of this assemblage. The potassic zones appear to have formed at high temperature (qualitatively in the range 580-390°C) in the presence of water in the magmatic isotopic range. This same study indicated that early hydrothermal biotite from Butte was formed from fluids containing meteoric waters. The conclusion that potassic zone minerals formed at high temperature in the presence of magmatic fluids appears justified but magmatic fluids may be diluted by meteoric water in some deposits.

Adjacent to the potassic zone (the potassium silicate zone of Meyer and Hemley, 1967) and gradational with it over a space of 30-90 m (100-300 ft) for many deposits is the phyllic zone, dominated by quartz, sericite, and pyrite (Lowell and Guilbert, 1970). Minor chlorite, rutile, clay minerals, and pyrophyllite commonly occur in the phyllic zone, but some deposits exhaustively studied (e.g., Butte) may be shown to contain many other minor minerals as well. Sericite predominates, however, as a fine-grained felted matlike pervasive replacement of other silicates and, in particular, as a coarser grained lining of quartz-sulfide veinlets. Vestiges of cleavage, zoning, and twin planes of silicates may be reflected in preferred orientation of sericite flecks. Primary quartz is generally overgrown with secondary quartz which is present beyond that normally expected by sericitization of other silicates. In some deposits (e.g., La Caridad) silicification in the phyllic zone makes that part of the deposit highly resistant to erosion. Pyrite prominently occurs as a dissemination of cubes within the sericitic groundmass. The virtual absence of carbonates and anhydrite from this zone is a striking feature in many deposits.

Sheppard, et al. (1971) and Taylor (1974) show a systematic correlation for D/H ratios in phyllic zone sericites for most Tertiary porphyry deposits and those deposits that require the presence of a meteoric water component in the hydrothermal fluid involved in sericitization. The O^{18} shift obtained in this study supports the conclusion that meteoric water contributed significantly to the hydrothermal fluid present during sericitization. Utilization of O^{18}/O^{16} geothermometry shows sericitization to have formed in the range 390-285°C, qualitatively speaking.

Peripheral to the phyllic zone and gradational from it for a variable distance is the argillic zone, which is characterized by either an illite-kaolinite or montmorillonite-kaolinite replacement of silicates, particularly plagioclase. Chlorite, pyrite, primary biotite, and minor sericite (lining fractures generally) also occur variably. Kaolin appears to dominate in many deposits. Sheppard, et al. (1971) and Taylor (1974) note that supergene clays have a greater O^{18} shift relative to hypogene. Studies to date indicate that meteoric water probably composed 50% or more of the hydrothermal fluid responsible for argillic zone hypogene alteration minerals. A maximum temperature of 300°C is assumed for kaolinization in the argillic zone.

Exterior to and gradational with the argillic zone over a distance of perhaps 30 m

(100 ft) is the propylitic zone. Chlorite is the characteristic mineral, but extensive dissemination of calcite, pyrite, epidote, and albite and variable amounts of montmorillonite, kaolin, and the other clays are present in this zone, which is the least distinctive and most widely distributed, fading gradually into fresh rock. Characteristic pervasive chloritization may coincide with the limit of pyritization or extend well beyond it.

Sheppard, et al. (1971) show vein carbonates to have slightly heavier C^{13} values than most primary sedimentary carbonates and to be consistent with a deep source. This study also provided a formation temperature between 150° and 210°C for vein calcite from Santa Rita, a temperature apparently projectable to formation of the propylitic zone (Sheppard, et al., 1971).

Data available in alteration zoning of porphyry copper deposits suggest the ore fluid traveled outward from a central heat source, starting at high temperatures. Paragenetically, then, the potassic zone is the earliest area altered. Successive zones around it are formed in turn with the propylitic last. As mineralizing fluids circulate through rock, the solids equilibrate with fluid-composition conditions. The change sequence begins with those characteristic of the propylitic and ends with those characteristic of the potassic. Interior zones, the earliest altered, develop at the expense of and hence prior to the next outer zone. Alteration zones have also been interpreted to form simultaneously as well as in sequence from potassic to propylitic.

Should any alteration zones be developed within nonigneous host rocks the mineral assemblages indicated to exist for each zone may be significantly modified by chemical composition of the host rock involved. The large extensive skarns present at Santa Rita, Cananea, and Bingham are the result of metasomatism of limy sediments within the porphyry copper alteration halo.

Supergene Alteration: A supergene alteration zone has been superimposed on pre-existing hypogene hydrothermal alteration at and near the surface for all porphyry copper deposits (Rose, 1970). Weathering is the principal factor in developing supergene alteration, which is the zone where cool descending oxygenated meteoric water may aid in oxidation and leaching of preexisting mineral assemblages.

Oxidation of ubiquitous pyrite and any other sulfides present in a porphyry copper system supplies hydrogen ion. Neutralization of the acid generated by hypogene silicates constitutes the greatest effect of supergene alteration. Hydrated silicates free of or deficient in alkali earths and metals are prominent in the supergene environment. Sulfates are ubiquitous. Kaolin is the most common secondary mineral thus formed, although conversion of biotite to a sericite-appearing mineral (Rose, 1970) and presence of other secondary silicates mask the original mineralogy. Supergene destruction of orthoclase and possible conversion of biotite and chlorite to sericite-appearing minerals at the surface could suggest that the potassic type alteration of Creasey (1959) may be missing near the surface when this is not the case. Unfortunately kaolin and other silicates formed by supergene processes also could have been formed by hypogene hydrothermal alteration and the final determination of which minerals are supergene and which hypogene is not always clear.

Sulfates formed by supergene processes are dominated by gypsum, alunite, and various complex hydrated ferric salts. These sulfates tend to occur mostly in fractures or joints with gypsum appearing erratically throughout the mineralized outcrop. In areas of stronger hypogene pyrite iron may be substantially but rarely completely leached from outcrop during weathering. Hydroxide and sulfate-bearing iron salts may appear around the iron-depleted zone, as is hypothetically shown in Fig. 44. The zones grade one into the other with no clear contact. Surrounding areas both of iron depletion and deposition is a zone where pyrite becomes less weathered. The distribution of iron minerals in capping depicted in Fig. 44 is idealized from a number of high pyrite deposits and does not represent outcrop zoning of all deposits in the Cordilleran orogen. Clearly those with

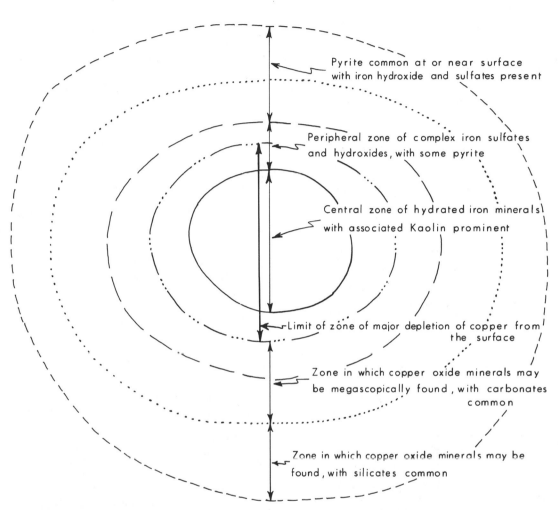

Pyrite common at or near surface
with iron hydroxide and sulfates present

Peripheral zone of complex iron sulfates
and hydroxides, with some pyrite

Central zone of hydrated iron minerals
with associated Kaolin prominent

Limit of zone of major depletion of copper from
the surface

Zone in which copper oxide minerals may
be megascopically found, with carbonates
common

Zone in which copper oxide minerals may be
found, with silicates common

Fig. 44. Oxide copper and iron minerals found around a high pyrite source. This figure analyses leached capping for a porphyry copper deposit having a hypogene high pyrite zone. Surface distribution of copper oxide minerals is shown around the high pyrite zone. Copper oxides including carbonates nearest the high pyrite zone and copper oxides including silicates away from the zone typify the leached cap. Copper will be substantially leached in the capping derived from the high sulfide zone. A change in iron oxide mineralogy is also detectable as distance is gained from the high sulfide center.

a below average pyrite content (e.g., Ajo) will provide a different surface distribution of oxidation products and types from those deposits with a high pyrite content (San Manuel).

A porphyry copper system with adequate hypogene pyrite and suitable channels for internal drainage may also be devoid of megascopically identifiable secondary copper minerals in weathered areas where hypogene pyrite was strongest. Leached capping in these areas may contain no more than 0.04% Cu although covering hypogene ore grading ten times this amount. Adjacent to this leached zone is an area where various copper oxide minerals occur at the surface. The type of surficial copper oxide minerals changes as distance is gained from the weathered area of strongest hypogene pyrite. Carbonates (azurite or malachite) tend to form near leached areas of strong hypogene sulfide, whereas copper silicates occur more commonly at some distance from high sulfur zones. Copper oxide minerals present with zonally arranged carbonates and silicates commonly include oxides and hydrated sulfates. The distribution of copper oxidation products shown in Fig. 44 depends not only on the presence and amount of pyrite but also on the reactivity of the rocks. Extensive supergene kaolinization is conducive to and indicative of greater mobility of both iron and copper.

Sulfates most commonly found other than those of copper and iron may be either supergene or hypogene. Anhydrite appears to be common to all hypogene porphyry copper ore systems (Meyer and Hemley 1967) and gypsum, its hydration product, is found in the capping of all porphyry copper deposits. At least some sulfate ion could have been derived from oxidizing sulfide, however. Alunite may also be supergene (Rose, 1970) or hypogene (Salas and Hollister, 1972).

Molybdenum in the supergene zone tends to be fixed during oxidation.

Mineralization

Fig. 41 shows the location of most generally known porphyry type deposits as well as the regional Bouguer gravity map (modified after Lyons, 1950). Negative anomalies coincide with a greater depth to the Moho, but porphyry copper deposits do not seem to occur in areas where thickest crust is inferred. Porphyry molybdenum deposits, on the other hand, are restricted to those areas where the crust is abnormally thick. Porphyry copper deposits and porphyry molybdenum deposits apparently do not occur in close proximity to one another. However, porphyry molybdenum deposits in Colorado have been dated as penecontemporaneous with some porphyry copper deposits in Utah and Nevada, where continental crust is thinner. The absence of porphyry copper deposits in areas of thick crust constitutes a notable exception to orogen-wide distribution. The present sialic crust is probably the crustal condition that prevailed during most periods of porphyry copper formation and therefore the configuration shown in Fig. 41 is meaningful.

As noted previously, the presence of economically significant molybdenite only in those porphyry copper deposits east of the western limit or Precambrian crust (Rogers, et al., 1974) suggests that crustal composition strongly influences geographic distribution of this element. Deposits found west of the western limit of the Precambrian (Rogers, et al., 1974) contain abnormally low molybdenum:copper ratios but commonly have high gold:copper ratios. The host for these latter occurrences includes a pre-middle Jurassic marine volcanic sequence. The metallogenic progression from thickest sialic crust (which contains porphyry molybdenum) to thinner crust (which contains porphyry copper-molybdenum) to thick accumulations of marine volcanic rocks (which host porphyry copper-gold) should not be ascribed to chance. The high positive correlation coefficient between copper-gold deposits to the west and copper-molybdenum deposits to the east of the western limit of Precambrian is statistically meaningful. The molybdenum and gold may be largely crustal derivatives. On the other hand presence of copper in both types of deposit on either side of the Rogers, et al. (1974) limit would infer that the crust is not a significant source factor for this metal. A substantial contribution of copper from a subcrustal (mantle) source seems required (Noble, 1974). A mantle source for some copper does not necessarily exclude contribution from a crustal source, however. In any case consistent variation of molybdenum with crustal thickness and composition infers a significant contribution of metals is made by the crust as porphyry copper hydrothermal solutions rise through it.

The character of hypogene sulfide mineralization has subtle distinctions in each alteration zone (Rose, 1970; Lowell and Guilbert, 1970). In the potassic zone, veins contain quartz-orthoclase-biotite-chalcopyrite-bornite (if present) or any combination including one or more of these minerals. Hypogene copper sulfide also occurs disseminated within potassic zone minerals, demonstrating paragenetic simultaneity of copper deposition and potassium metasomatism. Quartz-molybdenite-orthoclase and quartz-molybdenite biotite veinlets also occur in the potassic zone, indicating that at least some molybdenite mineralization also may be associated with potassic metasomatism. Gold as well as molybdenum may be concentrated in the potassic zone (e.g., Bingham).

If present in a porphyry copper system, bornite invariably occurs in the potassic zone. Pyrite only rarely is deposited simultaneously with bornite but if this occurs, pyrite does not form in close proximity to the bornite. On the other hand, examples of bornite occurring with magnetite are known. The chalcopyrite-pyrite and bornite-magnetite couples are well established in some deposits in the province. Zoning, with chalcopyrite appearing peripheral to a bornite core, is common.

Molybdenite, chalcopyrite, and bornite occur in the potassic zone as disseminations, apart from fracture filling. In many deposits disseminations of ore sulfides are not as economically significant as are sulfides filling fractures, although at Bingham dissemination makes up 50% of the ore values. Dissemination is largely restricted to the potassic zone and is not dependent upon presence of a centrally located stock. Ore sulfide dissemination can be seen in the potassic zone at Ray, for example, although no centrally located stock is present within this zone. In some deposits the potassic zone represents relatively weak metallization (e.g., Battle Mountain, Esperanza, Mineral Park). The central potassic zone at Santa Rita and San Manuel has a lower tenor than its gradational contact with the phyllic zone. A barren center may be present within the potassic zone (e.g., Cerrillos). In many deposits, on the other hand, the potassic zone is well metallized (e.g., Yerington, Bingham, Morenci, Bagdad, and Ajo) while the adjacent phyllic zone is poorly metallized or barren. Typically

the potassic zone has a low pyrite-chalcopyrite ratio and molybdenite (and bornite if present) tends to be most abundant in this zone.

Hypogene mineralization in the phyllic zone may be significant (e.g., Butte, Bingham, Castle Dome, Morenci) but is almost entirely restricted to fracture filling of quartz silicates and sulfides. Total sulfide content increases in the phyllic zone as distance is gained from the potassic zone until a maximum pyrite content is reached. Chalcopyrite is most strongly developed in a number of deposits at and near the interface of the potassic and phyllic zones. The pyrite-chalcopyrite ratio in veins and other fractures increases along with progressive increase in pyrite. Total hypogene copper content decreases toward the outer limits of the phyllic zone whereas total pyrite content regularly increases. Only rarely (as in Butte) is hypogene copper significant in the argillic zone. Molybdenite may be of subsidiary importance in the phyllic zone although important exceptions are known. In many deposits secondary silica is associated with molybdenite and this type of ore may occur in the phyllic zone rather than the potassic. The quasi-independent nature of molybdenum's distribution with respect to copper is well demonstrated in some deposits.

Gold also may be concentrated in the phyllic or argillic zone, as at Battle Mountain. This copper-gold deposit in an area Churkin (1974) shows to penetrate a thin crust has both phyllic zone gold-copper concentrations (Canyon ore zone) and external to these, argillic zone gold deposits. A thick marine volcanic section is projected to underlie the deposit. Gold may be identified with the potassic (Bingham), phyllic, or argillic (Battle Mountain) zones; hence it may be impossible to correlate its presence with any particular alteration assemblage.

Enargite is located in the upper levels of the complex porphyry system of Butte and has the same approximate distribution at Esperanza. It is a paragenetically young mineral most commonly found in the phyllic zone.

Tungsten (primarily wolframite) occurs in a number of Mexican porphyry copper occurrences in both the potassic and the phyllic zones. Although not recovered as a byproduct of molybdenum and copper in any existing operation the widespread (but very minor) level of wolframite mineralization distinguishes the Mexican province. Precambrian sialic crust probably underlies at a shallow depth those deposits with known tungsten mineralization.

Zinc and lead may occur significantly in either the phyllic or (more probably) the argillic and propylitic zones if a metallogenically zoned porphyry copper district is present. The phyllic zone commonly contains the same weight percentage of zinc as it does of molybdenum. Lead and zinc generally show greatest development in the propylitic zone or in rocks carrying an equivalent alteration if they are not igneous. Galena:sphalerite ratios increase from the phyllic zone to the propylitic zone. Both minerals are almost entirely confined to fracture filling with disseminations almost unknown. Isotopic studies at Bingham and Butte demonstrate a close identity between ore lead and lead in associated igneous rocks, permitting the conclusion that both had a common source. Contamination by crustal sources become more pronounced in leads located at a distance from the copper zone, however.

Generally, lead isotopic compositions within the porphyry system may not be those expected to result from leaching from the upper crust (Rose, 1970), further suggesting that whatever the source of the water or elements in the intrusion a portion of the metals and sulfur have a deep-seated common source.

S^{34}/S^{32} ratios of zero per mil (Field, 1966, and Rye and Ohmoto, 1974) for sulfur from sulfides in many porphyry copper deposits are consistent with origin from a homogenized crustal or mantle source. Rye and Ohmoto (1974) infer that porphyry copper with S^{34}/S^{32} values close to zero per mil is associated with a felsic pluton and the sulfur was probably derived from igneous sources. The igneous rock introduced the sulfur and appears to have introduced at least some metal as well.

A number of nonmutually supporting lines of evidence have been used in the past to postulate depths of erosion necessary to expose the porphyry copper deposits of the southern Cordilleran orogen (Lowell, 1974). On the other hand, Nash and Cunningham (1974) propose a well documented depth of burial based on fluid inclusion studies for the porphyry system now exposed at Bagdad and conclude that the water table surface at time of mineralization was approximately 1800 m (6000 ft) above present exposures. Nash and Theodore (1971) use fluid inclusion studies to show that the water table at Battle Mountain also was about 1800 m (6000 ft) above present surface at time of mineralization. A similar study at Bingham (Moore and Nash, 1974) suggests that the hydrostatic pressure for minerals now exposed was also 1800 m (6000 ft) at time of formation.

Supergene Enrichment: Supergene copper sulfide concentrations have been found associated with most major porphyry copper deposits in the southern Cordilleran orogen. Secondary copper sulfide enrichment results from oxidation of hypogene copper sulfide minerals, downward migration of water-soluble cupric ions in an acidic sulfate solution, and precipitation as copper sulfide by reaction with hypogene sulfide minerals. Examples of sulfide enrichment both above and below the water table are well known; hence using the water table as a plane to establish a controlling redox potential is unsound. Factors that have greatest influence on movement of copper in ground water appear to be:

1) Presence of sufficient pyrite or other sulfide to generate free hydrogen ion.

2) Absence of strong neutralizers in host rocks.

3) Adequate supply of ground water to provide movement.

4) Adequate internal drainage (e.g., through fractures and stockworks).

5) Sufficient copper in hypogene sulfides.

Once the copper is mobile, the most important factors influencing the precipitation of secondary sulfide (chalcocite or covellite) are the type of metal sulfide or sulfides present and their stability in an acid solution containing copper ion, the pH of the solution, and the redox potential.

Bacterial aids to the chemistry of solution and precipitation have been suggested as necessary adjuncts to processes forming supergene copper sulfide deposits. The physical chemistry of the system shows, however, that reactions will proceed whether bacteria are present or not. Anderson (1955) presents a clear summary of the chemistry involved as well as examples of supergene ores from the southern Cordilleran orogen.

Silver tends to concentrate in the supergene copper sulfide blanket. Silver:copper ratios may decrease slightly in the supergene zone but this is due to a relative flooding of more abundant copper.

Structure

Mineralization within southern Cordilleran orogen porphyry copper deposits occurs either in what is dominantly a breccia pipe or alternatively in a stockwork structure. The distinction in this area is less noticable than in the Andes because the occurrence of tourmaline as prominent black cementing mineral in Cordilleran deposit breccia matrix assemblages is less common. Also, small breccia pipes occur in nearly all stockwork deposits and stockworks occur in and close to all major breccia pipe deposits. Nevertheless, the categories stockwork and breccia pipe are included in Table 15 for those who wish to use them.

Although this generalized statement applies to the porphyry copper province as a whole as discussed in this chapter, a tourmaline-rich subprovince extends from Cananea south. Half the deposits shown in Fig. 41 south of Cananea include tourmaline within the alteration mineral assemblages. Most have identifiable tourmaline deposition in breccia pipes.

The distinction between breccia pipe and stockwork is still valid, however, for the following reasons:

1) Major breccia pipes (e.g., La Caridad) usually are surrounded by a circular or oval set of mineralized fractures the density of which may reach that of a stockwork.

2) Stockwork deposits have at least one and usually two major veinlet trends and breccia pipes associated with these are small and may be elongated in the direction of one veinlet trend. The major veinlet trend is commonly that of a strike-slip fault present within and well beyond the limit of the deposit.

3) Stockwork deposits frequently occur on or near an identifiable megastructure, whether this is actually a fault (e.g., Silver Bell), a linear arrangement of deposits (e.g., the San Pedro line), or a linear arrangement of deposits and alteration zones (e.g., the Cananea line).

Megastructures: Porphyry copper deposits tend to occur in northwest-trending groupings or lines (e.g., Walker line in the sense of Lowell, 1974, shown in Fig. 40) in the southern Cordilleran orogen. The deposits included in this map may not be a complete exposition, but enough are shown to adequately represent the spatial distribution of such deposits.

Megastructural lines display ages of activity that include dates found in individual deposits within the line. Fig. 40 shows four more commonly accepted northwest-trending lines discussed in the literature. From east to west, the northwest zones are the Battle Mountain (Roberts, 1966), Walker, San Pedro, and Cananea lines (Lowell, 1974). They appear remarkably parallel over much of the Cordilleran orogen. Where deposits of widely differing ages occur on a line, it has been claimed that the line was reactivated repeatedly over a long time span (e.g., Lowell, 1974, proposes a 140 m.y. time span for N40W-trending San Pedro line). If the Yerington deposit is located on the N40W-trending Walker line of Jerome and Cook (1967), the period of activity for this line extends intermittently from 111 m.y. rather than from the 17 m.y. suggested by Stewart (1971).

Hardyman (1975) proposes a possible Tertiary right slip separation in the northern

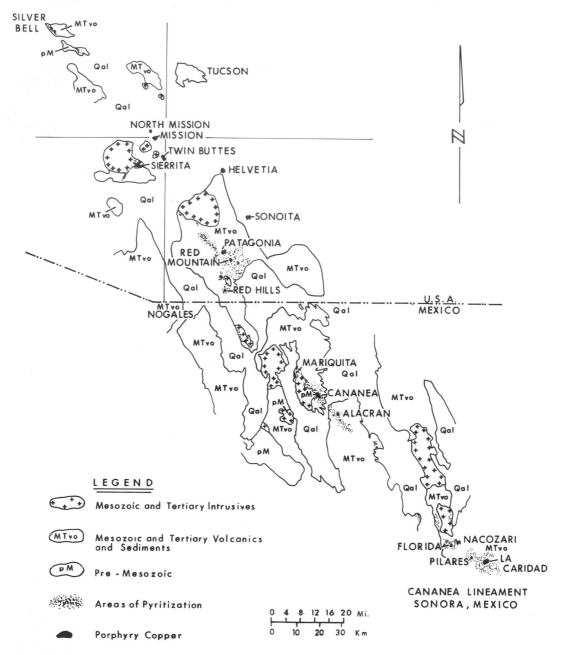

Fig. 45. Cananea line. From La Caridad to the vicinity of Tucson, discontinuous zones of pyrite dissemination occur between known porphyry copper deposits. The zones are elongated parallel to the trend of the line of deposits and add to continuity from one porphyry occurrence to the next. Additionally, pyrite halos around each porphyry copper are elongate along the line of deposits, giving substance to Cananea line as a lineal grouping of deposits from La Caridad northwest (*modified after Salas and Hollister, 1972*).

part of the Walker line. His evidence for displacement does not conflict with the concept that the Walker line is an old strike-slip fault that has been erratically active since the Mesozoic. The period of activity on the more clearly definable Cananea line was roughly from 53 m.y. for La Caridad to 63 m.y. for Arizona porphyry copper deposits.

Lowell (1974) and Jerome and Cook (1967) cite N60E lines of Laramide porphyry copper deposits in addition to the N40W lines mentioned previously.

The northeast-trending belts of porphyry copper deposits have also been cited by many in the past, although Noble (1970) points out that there may be no reflection of these lines in visible crustal structure.

Burnham (1959) and many others have speculated on the origin of these lines. A few have noticed other less conspicuous northwest-trending lines and have called attention to them. Fisher (1972) recognized the northwest-trending belt that includes Kerwin but did not name it, although Butte would project on such a line as would the Rocky Mountain Trench to the northwest.

The persistence for many miles of these lines along a strike and the evidence that activity occurred sporadically on them over a span of many years suggest that these zones should persist vertically and include fractures or fault segments capable of penetrating deeply into the crust or subcrust. The fracture may tap the mantle but at least it is capable of tapping a different and deeper melt or ore source than those fractures that localize other types of ore deposits. Existence of the linear features also suggest that the clustering of porphyry copper deposits around the Colorado plateau or on the border of an Arizona Cretaceous basin could be misinterpreted in the genesis of the deposits.

Stockwork Structures: Descriptions of detailed structure within a number of deposits presented by Rehrig and Heidrick (1972) indicate that stockwork trends in the majority of Arizona deposits reflect the two linear trends, N40W and N60E. Deposits falling on the Twin Buttes, Mission, Silver Bell, Ithica Peak (Mineral Park) line contain a set of stockwork that parallels these trends. A northeast-trending stockwork, which may be dominant, also exists in each deposit and ore occurs where the two trends intersect. Veinlets in the stockworks are steep to vertical and need not form a conjugate set. Additional conjugate mineralized fractures are present for each of the N40W or N60E major sets, probably giving rise to the frequently described crackle breccia, which consists of veinlets of three or more apparent trends.

The same trends for stockworks are present outside porphyry deposits in the Twin Buttes-Mineral Park line, however, and may be found throughout most of the southwest's stockwork deposits. The northeast set may vary from N20 to 60E and the northwest set may vary from N20 to 45W in any particular deposit, but persistence of these stockwork trends over a large portion of deposits in the southwest infers that the stockworks are not a response to autofractural tendencies within each magma (e.g., cooling cracks, volume change through heat loss) but to regional stress-strain relationships (Rehrig and Heidrick, 1976).

Most major porphyry copper deposits in the southern Cordilleran orogen occur on regional megastructures but not all have been associated with large strike-slip faults. Bingham is a fairly typical example of control of the emplacement of the composite intrusive, in part at least, by such a preexisting regional fault. This structure is reflected in the orientation of the veinlet trend within the ore body as well as in the distribution and orientation of late-stage dikes. The sericite zone peripheral to the potassic also appears to have been influenced by the regional fracture pattern (Moore and Nash, 1974).

Circular fracture patterns are also present in some deposits in response to intrusion of a specific pluton—El Arco, Fig. 46, for example.

Breccia Pipe: All stockwork deposits have some small breccia pipes associated with them. At Copper Creek near San Manuel (Fig. 41) the numerous small pipes present at the surface give way to a stockwork deposit 762 m (2500 ft) below the

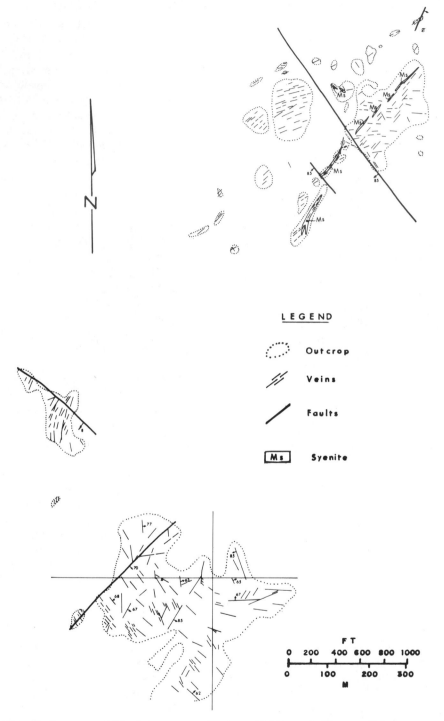

Fig. 46. Structure visible in outcrop at El Arco. Veinlet and dike trends in outcrop in the El Arco deposit form an incomplete arcuate set around an unmineralized center. Existence of a crossing set may be interpreted as a radiating group of fractures.

present surface. The Copper Basin (southeast of Bagdad, see Fig. 41) cluster of breccia pipes (Fig. 47) may infer a similar porphyry copper at depth.

In most other stockwork deposits now exposed these smaller breccia pipes appear to be scattered haphazardly over the ore body, occurring more frequently near the margins. Characteristics of these smaller pipes are distinct from those of larger pipes, however.

The larger pipes (e.g., La Caridad, Bisbee, Cananea) tend to be surrounded by a circular or oval fracture pattern. Northeast- and northwest-trending veinlets may be superimposed on the breccia and its circular zone but the pipe and its circular fracture zone are usually visible. The circular fracture zone is never perfectly developed as a continuous ring fault (Fig. 47) but rather is made up of discontinuous joints and faults parallel to the general outline of the pipe. Mineralization commonly occurs in these fractures, suggesting that these openings were available to the same mineralizing fluids as the pipe itself. As may be noted in Fig. 46, circular fractures may appear without the presence of a major pipe. The circular veinlet trend in Fig. 46 occurs around the El Arco pluton. The ore body consists of circular mineralized fractures. The weakly developed circular fracture pattern at Copper Basin (Fig. 47, after Vuich, 1970, and Johnston and Lowell, 1961) may reflect either a central nonoutcropping breccia pipe or a stock. A weakly developed northeast-trending set and an even fainter northwest-trending set are coexistent with the circular fracture pattern.

DISCUSSION

Accumulated evidence suggests that some elements of calc-alkalic stocks associated with porphyry copper deposits originated in the mantle and that in many deposits sulfide sulfur is also mantle derived. The linear arrangement of porphyry type mineralization along well defined zones (e.g., Cananea line) implies that a major fracture zone permitted or governed access to the surface or upper crust for at least some fraction of the magma and its metals. The near absence of porphyry

copper deposits in areas of thickest sialic crust may be coincidental but suggests that these areas would be unfavorable for copper exploration. This absence also suggests that fractures may penetrate thinner crust to permit esape from the mantle of the magma and its mineralizers to form porphyry copper deposits.

Another problem arises concerning minor metal values found within porphyry copper deposits of the southern Cordilleran orogen. Most are copper-molybdenum porphyries but minor gold occurs in all of them. In detail, porphyry occurrences found west of the western limit of the Precambrian have abnormally low molybdenum:copper but unusually high gold:copper ratios, relatively speaking. For deposits occurring east of the western limit of the Precambrian, Bisbee appears to be an important exception to the general rule that molybdenum significantly accompanies copper in this type of deposit. Bisbee also appears to contain an abnormally high (by Cordilleran standards) gold to copper ratio. Bisbee occurs in a trough with a very thick sedimentary column to the south, suggesting that the tectonic setting possibly is unusual. If rifting reduced the thickness of continental crust in its vicinity, it may be speculated that molybdenum impoverishment in this deposit reflects crustal setting, providing that molybdenum in porphyry copper deposits is supplied by the crust. The case for supplying copper from the mantle has already been made.

This raises the question why some stocks are barren while others have associated porphyry copper deposits. Speculations concerning this problem are fruitless without a much better understanding of the geometry of the crust and a more complete knowledge of the chemistry and physics of the lower crust and mantle. Subcrustal heterogeniety clearly exercises some control over mineralization (Noble, 1974). Probable derivation of the original magma and the sulfur of mineralized plutons from the mantle establishes its importance in deciding which intrusion is copper bearing.

It seems clear that regional distensional

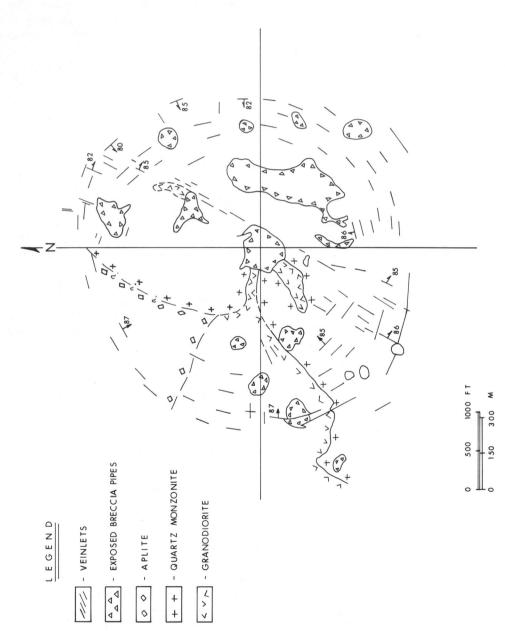

Fig. 47. Structural outline at Copper Basin. This figure shows generalized distribution and orientation of mineralized fractures in and near breccia pipes at Copper Basin, Ariz. The circular fracture pattern is superimposed on a northeasterly trending set of veins and veinlets.

or strike-slip tectonics favor development of porphyry copper deposits. At the time of porphyry copper formation distensional conditions within each deposit are reflected by the metals filling tensional fractures as hypogene sulfides.

Data in Table 15 and Fig. 40 also support the contention by Noble (1974) that subduction seems to bear no influence on the pattern of metal distribution or igneous activity. The process of formation of porphyry copper deposits in the southern Cordilleran orogen has no positive relation to subduction and is not governed by it (Lowell, 1974). The correlation between regional distension and porphyry copper formation does not seem to favor the derivation of metals or calc-alkalic magma from a subduction zone. Additionally, Noble (1974) points out that the space distribution of porphyry copper deposits lacks a significant pattern. These occur from Yerington to Kerwin and span nearly half the continent. A plate-consuming subduction zone acting as the unique source for copper could be expected to provide a more linear distribution, with deposits concentrated in a line equidistant from the subparallel to the coast. Furthermore, porphyry deposits appear to have random temporal distribution. Ely and Yerington are about 111 m.y. but Battle Mountain between them is 37 m.y. Bisbee is the oldest, dated 163 m.y., but the youngest is just to the north at 20 m.y. If porphyry copper deposits were related to subduction, then progressively younger ages eastward should be expected (Noble, 1974). This is clearly not the case for porphyry copper deposits. The age range of southern Cordilleran deposits from 163 m.y. to 20 m.y. suggests that they are not necessarily derived from subduction (Lowell, 1974), since the chronology of plate motion proposed in the past conflicts with these dates. Igneous rocks formed during periods of distension may include porphyry copper deposits, but those formed during periods of rapid subduction and batholithic development are nearly barren of such deposits.

The case for tying north-northwest trending fractures in a porphyry to north-northwest trending lines (Lowell, 1974) indicates that both the line and the fractures parallel the tectonic grain of the orogen. If tectonic grain is derived from plate motion (Burchfiel and Davis, 1972) porphyry copper deposits formed in fractures caused by plate motion. The parallel between the north-northwest lines of Fig. 40, the orogenic grain at time of mineralization, and the detailed structure within individual deposits is too great to have been caused by chance.

REFERENCES AND BIBLIOGRAPHY

Anderson, C. A., 1955, "Oxidation of Copper Sulfides and Secondary Enrichment," *Economic Geology,* Vol. 50, pp. 324-340.

Anderson, C. A., 1966, "Areal Geology of the Southwest," *Geology of the Porphyry Copper Deposits, Southwestern North America,* University of Arizona Press, Tucson, AZ.

Armstrong, R. L., and Suppe, J., 1973, "Potassium-Argon Geochronology of Mesozoic Igneous Rocks in Nevada, Utah, and Southern California," *Bulletin,* Geological Survey of America, Vol. 84, pp. 1375-1392.

Atwater, T., 1970, "Implications of Plate Tectonics for the Cenozoic Tectonics Evolution of Western North America," *Bulletin,* Geological Society of America, Vol. 81, pp. 3513-3536.

Barthelmy, D. A., 1974, "The El Arco Area, Baja California," *Abstracts,* Geological Society of America, Vol. 6, No. 3, p. 142.

Brooks, H. C., 1976, "Pre-Cenozoic Tectonic Framework, Oregon and Idaho," *Abstracts,* Geological Society of America, Vol. 8, No. 3, p. 357.

Burchfiel, B. C., and Davis, G. A., 1972, "Structural Framework and Evolution of the Southern Part of the Cordilleran Orogen, Western United States," *American Journal of Science,* Vol. 272, pp. 97-118.

Burnham, C. W., 1959, "Metallogenic Provinces of the Southwestern United States and Northern Mexico," Bulletin No. 65, New Mexico Bureau of Mines, p. 76.

Christiansen, R. L., and Lipman, 1972, "Cenozoic Volcanism and Plate Tectonic Evolution of Western United States," *Philosophical Transactions,* Royal Society of London, Series A, No. A-271.

Churkin, M., 1974, "Subcontinental Crust of the

Great Basin, Cordilleran Section," *Abstracts,* Geological Society of America, Vol. 6, p. 155.

Clark, K. F., 1972, "Stockwork Molybdenum Deposits in the Western Cordillera," *Economic Geology,* Vol. 67, p. 731.

Coney, P. J., 1972, "Cordilleran Tectonics and North American Plate Motion," *American Journal of Science,* Vol. 272, pp. 603-628.

Coney, P. J., 1975, "Overview of Lake Cretaceous and Cenozoic Plate Tectonics," *Abstracts,* Geological Society of America, Vol. 7, p. 1035.

Creasey, S. C., 1959, "Some Phase Relations in Hydrothermally Altered Rock of Porphyry Copper Deposits," *Economic Geology,* Vol. 54, p. 354.

Creasey, S. C., 1966, "Hydrothermal Alteration," *Geology of the Porphyry Copper Deposits, Southwestern North America,* University of Arizona Press, Tucson, AZ.

Drewes, H., 1976, "Laramide Tectonics from Paradise to Hells Gate," *Digest,* Arizona Geological Society, Vol. 10, p. 151.

Field, C. W., 1966, "Sulfur Isotope Data, Bingham," *Economic Geology,* Vol. 61, p. 850.

Fisher, F. S., 1972, "Tertiary Mineralization and Hydrothermal Alteration in the Stinkingwater Mining Region, Park County, Wyoming," Bulletin No. 1332, US Geological Survey.

Hardyman, R. F., 1975, "Cenozoic Faults in Northern Part of Walker Line, Nevada," *Abstracts,* Geological Society of America, Vol. 7, p. 1100.

Hedge, C. E., 1974, "Strontium Isotopes in Economic Geology," *Economic Geology,* Vol. 69, pp. 823-826.

James, A. H., 1971, "Hypothetical Diagrams of Several Porphyry Deposits," *Economic Geology,* Vol. 66, pp. 43-47.

Jerome, S. E., and Cook, D. R., 1967, "Relation of Some Metal Mining Districts in the Western United States to Regional Tectonic Environments and Igneous Activity," Bulletin No. 69, Nevada Bureau of Mines.

Johnston, W. P., and Lowell, J. D., 1961, "Geology and Origin of Mineralized Breccia Pipes, Copper Basin, Arizona," *Economic Geology,* Vol. 56, pp. 916-940.

Jones, D. L., Irwin, W. P., and Ovenshine, A. T., 1972, "Southwestern Alaska, A Displaced Continental Fragment," Professional Paper No. 800B, US Geological Survey, pp. 211-217.

Larson, R. L., and Pitman, W. C. III, 1972, "World-Wide Correlation of Mesozoic Magnetic Anomalies, and Its Implications," *Bulletin,* Geological Society of America, Vol. 83, pp. 3645-3662.

Lipman, P. W., Prostka, H. J., and Christiansen, R. L., 1971, "Evolving Subduction Zones in the Western United States, As Interpreted from Igneous Rocks," *Science,* Vol. 174, p. 882.

Livingston, D. E., Maugher, R. L., and Damon, P. E., 1968, "Geochronology of the Emplacement, Enrichment, and Preservation of Arizona Porphyry Copper Deposits," *Economic Geology,* Vol. 63, pp. 30-36.

Lowell, J. D., and Guilbert, J. M., 1970, "Lateral and Vertical Alteration-Mineralization Zoning in Porphyry Ore Deposits," *Economic Geology,* Vol. 65, No. 4.

Lowell, J. D., 1973, "Regional Characteristics of Southwestern North America Porphyry Copper Deposits, SME-AIME Preprint No. 73S12, Annual Meeting, Chicago.

Lowell, J. D., 1974, "Regional Characteristics of the Southwest," *Economic Geology,* Vol. 69, p. 601.

Lowell, J. D., 1976, "Trends and Techniques in Southwest Porphyry Exploration," *World Mining,* Vol. 34, p. 55.

Lyons, P. L., 1950, "A Gravity Map of the United States, *Digest,* Tulsa Geological Society, Vol. 18, p. 33.

Meyer, C., and Hemley, J. J., 1967, "Wallrock Alteration," *Geochemistry of Hydrothermal Ore Deposits,* Holt, Rinehart & Winston, New York, NY, p. 166.

Monger, J. W. H., Souther, J. G., and Gabrielse, H., 1972, "Evolution of the Canadian Cordillera: A Plate Tectonic Model," *American Journal of Science,* Vol. 272, pp. 577-602.

Moorbath, S., Hurley, P. M., and Fairbairn, H. W., 1967, "Evidence for Origin and Age from Sr-Rb Measurements," *Economic Geology,* Vol. 62, p. 228.

Moore, J. G., 1962, "K/Na Ratio of Cenozoic Igneous Rocks of the Western United States," *Geochemica et Cosmochimica Acta,* Vol. 26, 101-130.

Moore, W. J., and Nash, J. T., 1974, "Alteration and Fluid Inclusion Studies of the Porphyry Copper Body at Bingham, Utah," *Economic Geology,* Vol. 69, pp. 631-645.

Nash, J. T., and Theodore, T. G., 1971, "Ore Fluids in the Porphyry Copper Deposit at Copper Canyon, Nevada," *Economic Geology,* Vol. 66, p. 385.

Nash, J. T., and Cunningham, C. G., 1974, Fluid-Inclusion Studies of the Porphyry Copper Deposit at Bagdad," *Journal of Research,* US Geological Survey, Vol. 2, No. 1, p. 31.

Noble, D. C., 1972, "Some Observations on the Cenozoic Volcano-Tectonic Evolution of the Great Basin, Western United States," *Earth and Planetary Science,* Vol. 17, pp. 142-150.

Noble, J. A., 1974, "Metal Provinces and Metal Finding in the Western United States," *Mineralium Deposita,* Vol. 9, pp. 1-25.

Rehrig, W. A., and Heidrick, T. L., 1972, "Regional Fracturing in Laramide Stocks of Arizona, and Its Relationship to Porphyry Copper Mineralization," *Economic Geology,* Vol. 67, p. 184.

Rehrig, W. A., and Heidrick, T. L., 1976. "Regional Tectonic Stress During the Laramide and Late Tertiary, Basin and Range Province,"

Digest, Arizona Geological Society, Vol. 10, p. 205.

Richard, K., and Courtright, J. H., 1954, "Structure and Mineralization, Silver Bell," *Transactions,* AIME, Vol. 199, p. 1095.

Roberts, R. J., 1966, "Metallogenic Provinces and Mineral Belts in Nevada," Report No. 13, Nevada Bureau of Mines, pp. 47-72.

Rogers, J. J. W., et al., 1974, "Paleozoic and Lower Mesozoic Volcanism and Continental Growth in the Western United States," *Bulletin,* Geological Society of America, Vol. 85, pp. 1913-1924.

Rose, A. W., 1970, "Zonal Relations of Wallrock Alteration and Sulfide Distribution at Porphyry Copper Deposits," *Economic Geology,* Vol. 65, p. 920.

Rye, R. O., and Ohmoto, H., 1974, "Sulfur and Carbon Isotopes," *Economic Geology,* Vol. 69, pp. 826-842.

Salas, G., and Hollister, V. F., 1972, "Alteration Minerals as Ore Guides in the Porphyry Copper Province of Sonora, Mexico," *24th International Geological Congress,* Ottawa, Ont., Canada, Sec. 4, p. 261.

Schilling, J. H., 1971, "Miscellaneous K-Ar Ages of Nevada Intrusive Rocks," *Isochron/West,* No. 2, p. 46.

Schmitt, H. A., 1966, "The Porphyry Copper Deposits in Their Regional Setting," *Geology of the Porphyry Copper Deposits, Southwestern North America,* University of Arizona Press, Tucson, AZ.

Scholz, C. H., Baranzagni, M., and Sbar, M. L., 1971, "Late Cenozoic Evolution of the Great Basin," *Bulletin,* Geological Society of America, Vol. 82, pp. 2979-2990.

Sheppard, S. M. F. Nielson, R. L., and Taylor, H. P., 1971, "Hydrogen and Oxygen Isotope Ratios in Minerals from Porphyry Copper Deposits," *Economic Geology,* Vol. 66, p. 515.

Silver, L. T., and Anderson, T. H., 1974, "Possible Left-Lateral Disruption of the North American Craton Margin," *Abstracts,* Geological Society of America, Vol. 6, p. 955.

Stringham, B., 1966, "Igneous Rock Types and Host Rocks Associated with Porphyry Copper Deposits," *Geology of the Porphyry Copper Deposits, Southwestern North America,* University of Arizona Press, Tucson, AZ, p. 35.

Taylor, H. P., 1974, "The Application of Oxygen and Hydrogen Isotope Studies to Problems of Hydrothermal Alteration and Ore Deposition," *Economic Geology,* Vol. 69, pp. 843-883.

Vuich, J., 1970, Private Communication.

Zartman, R. E., 1974, "Lead Isotopic Provinces in the Cordillera of the Western United States and Their Geologic Significance," *Economic Geology,* Vol. 69, pp. 792-805.

6

Porphyry Copper Deposits of the Caribbean

CONTENTS

INTRODUCTION

Porphyry copper deposits have been reported from the Caribbean area by Pease (1966), Cox, et al. (1973), Guild (1974), Kesler, et al. (1975), and others. This chapter summarizes the most widely known characteristics of those deposits. Only one deposit that conceivably can be classed as a porphyry copper occurrence is in production (Pueblo Viejo, Dominican Republic), but others are known whose tonnage and grade are of economic interest. Deposits in this province may be classified as copper-gold or copper-molybdenum types (Kesler, 1972), but deposits associated with quartz-rich plutons that have high $Na_2O:K_2O$ and high gold:copper but low molybdenum:copper ratios dominate.

Porphyry copper occurrences are reported in the literature from Puerto Rico, Haiti, the Dominican Republic, Jamaica, and Cuba. Only the porphyry copper deposits from Puerto Rico are extensively described in the literature; consequently the Caribbean porphyry copper province in general is not as well documented as the southern Cordilleran orogen. Because past published technical descriptions are incomplete, some data presented here are new to the literature.

The Caribbean includes the island arcs between North and South America. Porphyry copper occurrences are known from the Greater Antilles, the generally east-west trending group of islands that lies south of North America. Fig. 48, therefore, only covers the Greater Antilles portion of the Caribbean.

Porphyry copper deposits in this area are not described by the Lowell and Guilbert (1970) model in details of metallization (general paucity of molybdenum), alteration (scarcity of orthoclase), or petrography (preponderance of quartz diorite). Therefore, this model should be modified to include deposits associated with quartz-bearing plutons such as those found in the Caribbean.

In the Meme deposit on Haiti ore occurs in skarn adjacent to the quartz diorite phase of a composite quartz monzonite-granodiorite-quartz diorite pluton (Kesler, 1968). This deposit, with a K-Ar 66 m.y. date, is not included in Fig. 48 as a porphyry copper deposit because metallization is restricted largely to the intruded rocks. Although the intrusive rocks at the Meme deposit may contain sulfide near skarn contacts they are otherwise essentially unaltered.

Porphyry molybdenum deposits as defined by Clarke (1972) have not been reported in the Caribbean area.

149

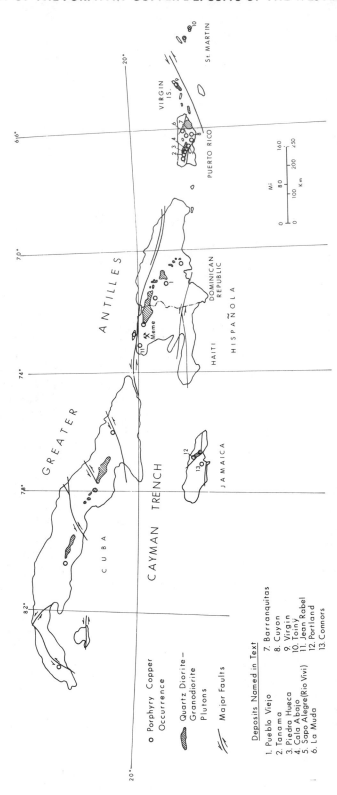

Fig. 48. Porphyry copper occurrences of the Greater Antilles. Major quartz-bearing plutons, strike-slip faults, and porphyry copper occurrences are shown on this map. Deposits in Puerto Rico and Jamaica are generalized, since the scale does not permit showing those closely clustered together. The Connors deposit includes Camel Hill, Mab, and Geo Hill, for example.

GEOLOGIC SETTING

Geologic History: The oldest Greater Antillean rocks are metamorphosed gray-wacke, argillite, tuff, mafic igneous rock, and carbonates of pre-Middle Jurassic age. These occur in west and central Cuba and western Jamaica. The oldest dated rocks, also found in Cuba, are terrigenous clastics and evaporites of Early and Middle Jurassic age. The island arc that hosts known porphyry copper occurrences developed from these older rocks eastward to San Martin, commencing in the Upper Jurassic and continuing to the Eocene. Locally 10 to 12 km of mafic to silicic igneous rocks accumulated (Khudoly and Meyerhoff, 1971). Except for western Cuba and Jamaica(?) the arc developed on a base of older oceanic crust.

Kesler, et al. (1975) note that the Mesozoic island-arc accumulations contain the ingredients of oceanic crust including ultramafic material. A Cretaceous trough faced the western Jamaican backland that Khudoly and Meyerhoff (1971) designated Paleozoic(?). Cretaceous volcanic rocks were deposited elsewhere in the Greater Antilles. Since Upper Cretaceous time compression and volcanism typical of island-arc development have given way to strike-slip fault tectonics. All porphyry copper deposits may eventually be demonstrated to have formed within this period of strike-slip tectonics, in large part accompanying plutons intruded in that interval. Figs. 48 and 49 (Cox and Briggs, 1973) show major strike-slip faults and porphyry copper occurrences. Except for

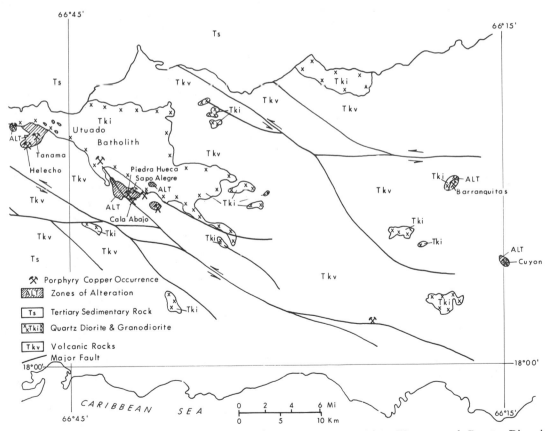

Fig. 49. Geologic outline and porphyry copper deposits of Puerto Rico. West central Puerto Rico is depicted on this map which shows quartz-bearing plutons, altered areas, and porphyry occurrences (*after Cox and Briggs, 1973*).

Puerto Rican deposits (Pease, 1966) the literature records only occasional involvement of porphyry deposits in the faults. However, detailed mapping on the other islands may show a genetic tie between these faults and the porphyry deposits. The faults at least represent the type of tectonic environment that favors porphyry development.

Plate Tectonic Relationships: Relative motion between the North and South American plates began at 180 m.y. (Ladd, 1973). According to the Ladd (1973) analysis of magnetic data in the North and South Atlantic, development of the Greater Antilles island arc coincided with a westerly movement (with either a northerly or southerly vector also present) of North American relative to South America from 180 to 85 m.y. From 53 to 9 m.y. the South American plate moved westerly (again with a minor northerly or southerly vector also present) relative to North America and the period of strike-slip tectonics in which most porphyry copper deposits developed stems from the 85 m.y. date. Radiometric dating on the deposits coincidentally shows all to be less than 85 m.y. old.

The Caribbean plate acted independently of the North and South American plates. Hess (1938) postulated left lateral and right lateral strike-slip movement along what are now considered its northern and southern boundaries.

Hispanola and Puerto Rico lie along the north side of the Caribbean plate and strike-slip movement continued on faults north of these islands from the Cretaceous (Cox, 1973). Because movement is north block west, displacement is sympathetic with plate movements discerned by Ladd (1973). Faults with strike-slip displacements north of Puerto Rico and Hispanola continue into the Cayman Trench.

Moore and Del Castillo (1974, p. 612) infer a slightly different model for development of the Antilles but concur with others that a strike-slip tectonic regime began developing at 85 m.y. and continued through the period when porphyry copper deposits developed in the deactivated island arc. From the point of view of porphyry copper genesis, concurrence by most students of the Caribbean that Tertiary strike-slip tectonics developed in a preexisting arc environment minimizes speculation concerning crustal and tectonic setting for these deposits.

PORPHYRY COPPER DEPOSITS

Petrography: Table 16 summarizes those intrusive phase compositions in each district closest to ore spatially and temporally. Potassium is present in sufficient volume in only a few deposits to raise the orthoclase:plagioclase ratio above the quartz diorite field. Most porphyry copper deposits are integral parts of plutonic complexes, and in all cases the quartz diorite phases are dominant (Kesler, et al., 1975). Plutons with high Na_2O: K_2O ratios are characteristic. Where granodiorite and quartz monzonite are found, orthoclase occurs as an intergranular mineral with quartz and between quartz, plagioclase, and ferromagnesian silicate phenocrysts. Plagioclase is nearly always zoned with an average composition of An_{40}. Biotite, hornblende, or both may occur as phenocrysts set in the microaplitic groundmass. Jamaican examples tend to be richer in potassium minerals than those of Puerto Rico and the Dominican Republic.

Zoned complexes (e.g., Utuado batholith, Fig. 49) with quartz monzonite and granodiorite cores ringed by quartz diorite may be spatially associated with some porphyry copper deposits (e.g., Tanama, Piedra Hueca, Cala Abajo, Sapo Alegre with the Utuado batholith, and Puerto Maunabo with the San Lorenzo batholith). Most deposits, however, are associated with smaller plutonic centers.

Regional Alteration: Large areas of hydrothermal alteration occur in the pre-Tertiary volcanic rocks of the Greater Antilles (Bergey, 1966). These are well documented only on Puerto Rico (Cox and Briggs, 1973) but they occur on other islands as well. Some of these alteration zones include porphyry copper deposits but others have been explored in vain for such occurrences. Propylitic-zone minerals most commonly are devel-

Table 16. Porphyry Copper Deposits of the Caribbean

Name	Place	Age, m.y.	Rock Intruded	Pluton	Alteration Zoning from Core	Metals	Pyrite Zone km x 10³ size	% in Phyllic Zone	References
Pueblo Viejo	Dominican Rep.	—	M Sed	Qtz Dio Por	Phy Arg Prop	Cu-Au	3 x 5	8	Guild (1975)
Tanama	Puerto Rico	62(?)	M Vol	Qtz Dio Por	Pot Phy Arg Prop	Cu-Au	4 x 6	5	Cox (1973)
Piedra Hueca	Puerto Rico	62(?)	M Vol	Qtz Dio Por	Pot Phy Arg Prop	Cu-Au	—	3	Cox (1973)
Cala Abajo	Puerto Rico	62(?)	M Vol	Qtz Dio Por	Pot Phy Arg Prop	Cu-Au	3 x 6	5	Cox (1973)
Sapo Alegre	Puerto Rico	41(?)	M Vol	Qtz Dio Por	Pot Phy Prop	Cu-Mo	—	3	Cox, et al. (1975)
La Muda	Puerto Rico	60	M Vol	Qtz Dio Por	Phy Arg Prop	Cu-Mo	2 x 3	3	Cox (1973)
Barranquitas	Puerto Rico	—	M Vol	Qtz Dio Por	Pot Phy Arg Prop	Cu-Au	2 x 3	4	Cox (1973)
Cuyon	Puerto Rico	47	M Vol	Qtz Dio Por	Pot Phy Arg Prop	Cu-Au	2 x 2	4	Cox (1973)
Virgin	Virgin Islands	31	M Vol	Grdr Por	Phy Arg Prop	Cu-Au	N.E.	6	Kesler, et al. (1975)
Toiny	St. Bartheleme	—	T Sed	Qtz Dio Por	Pot Phy Arg Prop	Cu-Au	2 x 2	4	Kesler, et al. (1975)
Jean Rabel	Haiti	—	M Vol	Qtz Dio Por	Phy Arg Prop	Cu-Au	2 x 3	6	Kesler, et al. (1975)
Portland	Jamaica	63	M Vol	Qtz Mon Por	Pot Phy Arg Prop	Cu-Au	2 x 4	6	—

Qtz: Quartz
Dio: Diorite
Grdr: Granodiorite

Por: Porphyry
Mon: Monzonite
M: Mesozoic
Vol: Volcanic rock

N.E.: Not exposed
Pot: Potassic
Phy: Phyllic

Arg: Argillic
Prop: Propylitic
T: Tertiary
Sed: Sedimentary rock

oped in these zones (Cox, et al., 1973). Although tinted green by abundant chlorite, the propylitic alteration assemblage also includes sericite, epidote, albite, calcite, dolomite, siderite, ankerite, rutile, and pyrite. Primary mafic minerals are completely altered to clusters of one or more minerals while plagioclase is variably altered to albite with or without sericite, epidote, calcite, and K-feldspar. Albite occurs pervasively as masses, veins, or disseminations.

Alteration Zoning: Porphyry copper deposits in the Caribbean contain all or some alteration zones exposed in deposits with quartz-bearing calc-alkalic plutons in other orogens. For conformity with other deposits the names potassic, phyllic, agrillic, and propylitic are retained. The mineralogy of each zone in Caribbean deposits commonly is distinct, however, from such zones found in deposits in the Andes, the Appalachian, or the Cordilleran orogen.

Central to the alteration zones may be a fresh core intrusion or a zone in which amphibole (hornblende or actinolite) may be prominent (as at Barranquitas and Tanama). Secondary amphibole is rarely noted in deposits in other orogens (e.g., in an occasional veinlet of secondary hornblende in the Highland Valley, the Babine Lake area of British Columbia, or Bingham) and occurs too infrequently to warrant inclusion in the alteration sequence of this model. At Barranquitas K-feldspar occurs as fine anhedral grains with amphibole.

Cox, et al. (1973) describe mineralogy for the potassic zone that includes secondary biotite, chlorite, epidote, or some combination of these with quartz but commonly without potassium feldspar. Chlorite seems to follow biotite paragenetically but does not occur zonally outward from it. Both are found as veins, irregular patches, and felted masses. Epidote in the potassic zone is younger than biotite and may be younger than chlorite. Table 16 indicates that 8 of the 12 deposits well enough documented for inclusion in the table have a potassic zone and Cox, et al. (1973) suggest that the potassic zones of half the Puerto Rican de-

posits may be devoid of K-feldspar. Microcline has been identified only at Piedra Hueca. Biotite is suggested to be potassium-bearing, however, because its alteration products (e.g., vermiculite) are accompanied by fine-grained intergrowths of orthoclase. Retention of the name potassic zone is therefore considered appropriate.

The potassic zone is core to a phyllic zone that contains characteristic quartz-sericite-pyrite dominant assemblages but that also may contain sericite-albite-pyrite (e.g., Sapo Alegre, Cox, et al., 1975) and sericite-chlorite-pyrite mixtures. The sericite-dominant alteration assemblages may contain epidote and kaolinite in addition to the chlorite, albite, pyrite, and quartz previously mentioned.

Hypogene argillic zone minerals characteristically develop erratically adjacent to the porphyry copper deposits of the Caribbean. The Pueblo Viejo deposit contains a large well developed argillic zone in intruded argillite, which is the locus of the 17.7 million mt of developed ore grading 0.25% Cu, 3.7 g (0.13 oz) Au, 31.8 g (1.12 oz) Ag, and 2.19% Zn (Anon., 1974). This zone is characterized by kaolinite, montmorillionite, chlorite, and pyrite as pervasive secondary minerals with illite and montmorillonite each locally prominent (Fig. 50). Cala Abajo and Tanama also contain large volumes of hypogene argillic zone mineral assemblages, primarily kaolin-pyrite-montmorillonite mixtures with prominent areas of hypogene antlerite, chlorite, albite, and quartz intermixed locally and erratically. Argillic zones in the remaining deposits shown in Table 16 are largely restricted to shear zones penetrating the propylitic zones containing a clay mineralogy similar to that of the larger zones.

Very large propylitic zones surround the potassic-phyllic-argillic zones in all Caribbean deposits. Fig. 49 shows the lateral extent of some Puerto Rican alteration zones associated with porphyry deposits.

All the alteration zones reflect the high $Na_2O:K_2O$ ratios present in most volcanic and plutonic rocks of this region. The absence

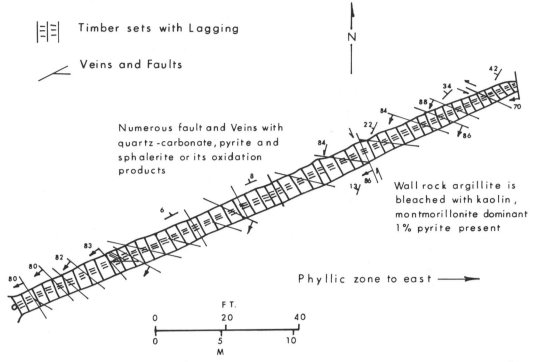

Fig. 50. Structure visible at Pueblo Viejo, Dominican Republic. Detailed geology of the No. 6 adit at Pueblo Viejo showing structure in the argillic zone is given in this figure. Veinlets contain sufficient gold, copper, and zinc to make this fringe portion of the porphyry deposit economic. Vein trends continue to the porphyry copper phyllic zone east of this adit.

or scarcity of orthoclase in the potassic zone is analogous to the mineral assemblage of this zone noted in some Cascades deposits as well as in the majority of deposits in the continental margin belt of Alaska. Common appearance of albite in most Caribbean alteration halos reflects the host-rock chemistry. Albite may occur in the phyllic zone (e.g., Sapo Alegre), in the argillic zone (e.g., Pueblo Viejo), and in the propylitic zone (e.g., Tanama) and therefore is not a guide mineral for copper.

Except for prominence of albite and impoverishment of orthoclase the mineral assemblages for each alteration zone are similar to those found in such zones of other orogens, so the names for these zones should be retained.

Mineralization: With very few exceptions mineralization found in Antilles porphyry copper deposits is the copper-gold type. Most deposits have low molybdenum:copper ratios. The most significant molybdenum content occurs in Jamaica where a possibly Paleozoic(?) basement may exist. Where molybdenum occurs significantly in Puerto Rico field evidence suggests the possibility of magmatic differentiation with molybdenum appearing contemporaneously with or prior to a magmatic phase enriched in potassium. The postmineral granodiorite at Sapo Alegre qualifies as the late differentiate in that deposit.

Gold is relatively more important in Caribbean deposits than in other regions discussed in this volume. For most examples, however, excepting only a few gold-rich bornite-bearing potassic zones, gold content for ores in potassic alteration zones is less important than in phyllic zone ores. The gold may be brought in with copper during intrusion of a mineralized pluton and then transported outward by highly saline magmatic-hydrothermal fluids. Alternatively, meteoric-hydrothermal fluids may have concentrated gold in the phyllic zone, leaching the element from the intruded basic marine volcanic rocks that host the system. For most deposits it is conceivable that both processes were important, although continued relative improvement of gold: copper ratios in the argillic zone of some deposits (e.g., Pueblo Viejo) is compatible with meteoric-hydrothermal transport. Gold is inconsequential in the propylitic zone of most porphyry deposits.

Metal zoning in most Greater Antilles deposits is discernible with a magnetite-bornite(if present)-chalcopyrite assemblage coexisting with the potassic zone. Metallic sulfide minerals may occur as disseminations in this zone or as quartz-sulfide-oxide veinlets or both. Pyrite-chalcopyrite-hematite(if present) occurs in the phyllic zone almost entirely as veinlets with quartz or as matrix cementing breccia fragments. Pyrite-chalcopyrite and sphalerite may occur in the argillic zone entirely as fracture filling with quartz or quartz-carbonate veins. Galena and sphalerite with minor chalcopyrite may occur as vein fillings in the propylitic zone with magnetite present in a few examples.

Structure: Except for well developed breccia pipes in the Portland deposit most Caribbean porphyry coppers are primarily stockworks. The dominant veinlet trend noted in each deposit is parallel or sub-parallel to the regional tectonic grain and also parallels major strike-slip faults existing nearby. Fig. 50 (Pueblo Viejo) exemplifies the type of veining found. This example is used because Pueblo Viejo is the only porphyry deposit in production and the structure is easily observed.

Stockwork veinlets reflect metasomatism as well as metallogeny in the alteration zone where they are found. Veinlets of the potassic zone not only contain chalcopyrite, quartz, anhydrite, and magnetite, but albite, chlorite, biotite, or epidote as well. The same set of veinlets in the phyllic zone may contain albite, sericite, quartz, and pyrite, whereas in the propylitic zone carbonates, chlorite, and albite are common in vein structures.

Age of the Deposits: Table 16 summarizes age dating on the best known deposits. The Connor deposit on Jamaica (not tabulated) has a K-Ar date of 85 m.y. (Kesler. et al., 1975). All the others are Tertiary. It is unlikely that any deposits will be found with ages older than 85 m.y. if correlation of tectonic regime and porphyry copper deposition is valid.

CONCLUSIONS

Porphyry copper deposits of the Caribbean developed in the period from 85 to 9 m.y. in a deactivated island-arc environment that included oceanic crust as a foundation. Paleozoic sialic crust may be suspected only in the western parts of Jamaica and Cuba. Except for a few rare small deposits these are copper-gold type porphyries. Their petrography is high sodium:potassium compared with other deposits developed in a cratonic setting. Alteration zoning reflects high soda content but the dominant secondary silicates developed (except for the absence of orthoclase) are still typical of the Lowell and Guilbert (1970) model, which should be expanded to incorporate these quartz-rich calc-alkalic deposits.

Distribution of copper and gold suggests important involvement of copper with a magmatic-hydrothermal fluid but leaves open a possible dependence of gold on meteoric-hydrothermal processes.

REFERENCES AND BIBLIOGRAPHY

Anon., 1974, "Pueblo Viejo," *Mining Magazine,* No. 1, p. 5.

Bergey, W. R., 1966, "Geochemical Prospecting for Copper in Puerto Rico," 3rd Caribbean Geological Conference, Geological Survey of Jamaica.

Clark, K. F., 1972, "Stockwork Molybdenum Deposits in the Western Cordillera," *Economic Geology,* Vol. 67, p. 731.

Cox, D. P., 1973, "Porphyry Copper Deposits of Puerto Rico and Their Relation to Arc-Trench Tectonics," Open File Report, US Geological Survey.

Cox, D. P., and Briggs, R. P., 1973, "Metallogenic Map of Puerto Rico," Map No. I-721, US Geological Survey.

Cox, D. P., Larson, R. R., and Tripp, R. B., 1973, "Hydrothermal Alteration in Puerto Rican Porphyry Copper Deposits," *Economic Geology,* Vol. 68, p. 1329.

Cox, D. P., Peres, G. I., and Nash, J. T., 1975, "Geology, Geochemistry, and Fluid-Inclusion Petrography of the Sapo Alegre Porphyry Copper Prospect and Its Metavolcanic Wall-Rocks, Puerto Rico," *Journal of Research,* US Geological Survey, Vol. 3, p. 313.

Guild, P. W., 1974, "Mineral Resources of the Caribbean Region," Open File Report, US Geological Survey.

Hess, H. H., 1938, "Gravity Anomalies and Island Arc Structure with Particular Reference to the West Indies," *Proceedings,* American Philosophical Society, Vol. 79, p. 71.

Kesler, S. E., 1968, "Contact Localized Ore Formation at the Meme Mine, Haiti," *Economic Geology,* Vol. 63, p. 541.

Kesler, S. E., 1972, "Copper, Molybdenum, and Gold Abundances in Porphyry Copper Deposits," *Economic Geology,* Vol. 67, p. 106.

Kesler, S. E., Jones, L. M., and Walker, R. L., 1975, "Intrusive Rocks Associated with Porphyry Copper Mineralization in Island Arc Area," *Economic Geology,* Vol. 70, p. 515.

Khudoley, K. M., and Meyerhoff, A. A., 1971, "Paleogeography and Geologic History of the Greater Antilles," Memoir No. 129, Geological Society of America.

Ladd, J. W., 1973, "Relative Motion Between North and South America and the Evolution of the Caribbean," *Abstracts,* Geological Society of America, Vol. 5, p. 705.

Lowell, J. D., and Guilbert, J. M., 1970, "Lateral and Vertical Alteration-Mineralization Zoning in Porphyry Copper Deposits," *Economic Geology,* Vol. 65, pp. 373-408.

Moore, G. W., and Del Castillo, L., 1974, "Tectonic Evolution of the Southern Gulf of Mexico," *Bulletin,* Geological Society of America, Vol. 85, pp. 607-618.

Pease, M. H., 1966, "Some Characteristics of Copper Mineralization in Puerto Rico," 3rd Caribbean Geological Conference, Geological Survey of Jamaica.

7

Porphyry Molybdenum Deposits
of the North American Cordillera

CONTENTS

INTRODUCTION

Within the past few years Clark (1972), King (1970), and King, et al. (1973), have summarized porphyry molybdenum occurrences. Clark (1972) includes as porphyry or stockwork deposits some occurrences more recent exploration has shown to be of other genetic types. King (1970) and King, et al. (1973) provide broad general descriptions that unfortunately are restricted to the US. Nor does the King, et al. (1973) description have benefit of the Wallace (1974), Hall, et al. (1974), or Giles (1975) data on alteration, fluid inclusions, or isotope chemistry. The synthesis presented here updates these older summaries, eliminating all but the well established porphyry molybdenum deposits as well as enumerating the unifying characteristics displayed by these deposits within the Cordillera of North America. This type of deposit occurs in the South American Andes as well, but Andean deposits are not well documented so only North American Cordilleran deposits are considered.

Stockwork, porphyry, and Climax type molybdenum deposits are synonymous geological labels for the large molybdenite deposits found in parts of the western Cordillera of North America (Khruschov, 1959). These labels describe and define the type of deposit discussed herein. The term porphyry is preferred for this chapter because it describes the common characteristics; it separates the molybdenum deposits from the porphyry copper deposits, which have many similarities but also important differences. This usage also eliminates blanket application of the word stockwork to breccia pipe deposits such as Boss Mountain in British Columbia.

J. R. Woodcock has contributed much to this chapter; his influence, original data, and thoughtful comments qualify him as co-author.

Porphyry molybdenum deposits are characterized by fine-grained molybdenite occurring alone or with quartz and/or pyrite in fractures and open spaces of stockworks and breccias. The molybdenite is generally less than one millimeter across and in many cases too fine-grained to be seen with the aid of a hand lens. Occasionally some deposits contain molybdenite crystals as large as one centimeter across; however, areas with such coarse-grained molybdenite are seldom economic. Porphyry molybdenite deposits are generally spatially and genetically associated with intrusions. Most plutons are of acid to intermediate rock with a porphyritic texture and fine-grained phaneritic to aphanitic matrix.

Although porphyry molybdenite deposits resemble some porphyry copper deposits in many aspects, especially in their relationship to intrusive stocks and association with large zones of intense hydrothermal alteration, there are some distinct differences. A log-log graph of copper content vs. molybdenum content for porphyry copper deposits and stockwork molybdenite deposits of North America show distinct separation into two populations with very few deposits having transitional grades.

Outstanding differences also exist for the ratio between other metals and molybdenum. The tungsten:molybdenum content for stockwork molybdenite deposits is much greater than that for most porphyry copper deposits. The rhenium:molybdenum ratio for stockwork molybdenite deposits is less than one-fifth that for porphyry copper deposits. Molybdenum deposits are characterized by fluorine-bearing minerals, either silicates (as

in a greisen) or the mineral fluorite. Fluorine minerals are less prominent in porphyry copper deposits. Additional differences will become evident from the following description of the molybdenite deposits.

Table 17 lists the principal well explored porphyry molybdenum deposits in the Cordilleran orogen, and Fig. 51 shows their location.

LOCATION OF THE DEPOSITS

The principal molybdenum deposits of the Cordilleran orogen are shown in Fig. 17. Except for Opodepe in Mexico these deposits form a northwest-trending band from Questa on the south to Adanac on the north. They range in age from 23 m.y. (Questa) to 141 m.y. (Endako) with no systematic trend apparent in space distribution of deposits with time. The apparent northwest trend is established by a variety of geologic phenomena not including control by a single megashear.

Regional Controls: The largest molybdenum deposits occur within the -200 Bouguer gravity contour (Fig. 51). Those that occur outside this contour contain less metal, which is frequently reflected in lower grade and diminished reserve. This is not to say that small deposits will not be found within the -200 contour; rather, the inference is clear that deposits external to the -200

Table 17. Significant Stockwork Molybdenum Deposits of the Cordilleran Orogen

| Deposit, Location | Age, m.y. | Reserve | | Reference |
		Million mt (million tons)	% Molybdenum	
Adanac (BC)	62	94.3 (104)	@ 0.09	White (1976)
BC Moly (BC)	54	97.9 (130)	@ 0.10	Giles (1975)
Hudson Bay Mountain (BC)	69	90.7 (100)	@ 0.12	Kirkham (1969)
Endako (BC)	141	168.7 (186)	@ 0.10	Drummond & Kimura (1969)
Boss Mountain (BC)	102	14.5 (16)	@ 0.30	Soregaroli (1975)
Urad-Henderson (CO)	26	272.1 (300)	@ 0.30	Wallace (1975)
Climax (CO)	32	208.7 (230)	@ 0.20	Hall, Friedman, & Nash (1974)
Questa (NM)	23	226.8 (250)	@ 0.11	Schilling (1956)
Big Ben (MT)	?	18.1 (20)	@ 0.20	Creasey & Scholz (1945)
Thompson Creek (ID)	86	90.7 (100)	@ 0.15	Anon. (1974)
Opodepe (Mexico)	?	136.1 (150)	@ 0.08	UN (1970)
Red Mountain (CO)	?	?	?	Thompson (1975), King (1970)
White Cloud—Boulder Creek (ID)	61	?	?	Ross (1937), Bennett (1973)
Wilson Arm—Quartz Hill (AK)	26	90.7 (100)	@ 0.15	Anon. (1976)
Cannivan (MT)	59	185.1 (204)	@ 0.096	Worthington (1977)

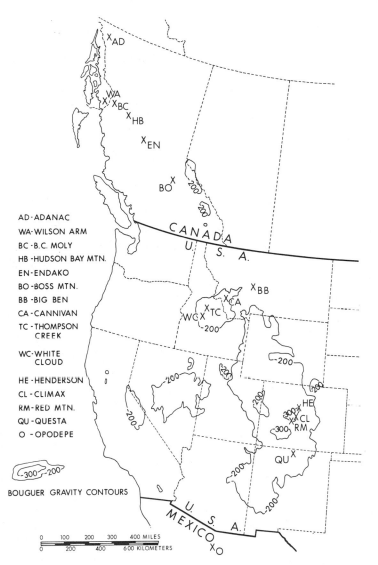

Fig. 51. Location of major porphyry molybdenum ore bodies now known. Northeast-trending clusters occur within the general northwest trend for all deposits. The largest deposits are found within the -200 Bouguer contour (*modified after Armstrong, et al., in press*).

contour are likely to be less attractive economically. Because the Bouguer contours reflect depth to the Moho the thickness of sialic crust appears to be a factor in metallization of this type of deposit.

No economically significant molybdenum deposits have been found to penetrate a thick section of marine andesitic or basic volcanic rock. All those listed in Table 17 and shown in Fig. 51 penetrate either schists and gneisses derived from sialic metamorphic rocks or intrude large differentiated quartz-bearing plutons. Thus the parameters controlling location of the deposits apparently include not only the thickness of the crust but its chemical composition as well.

In some areas deposits occur in belts or elongate clusters as in the northeast-trending Colorado Mineral Belt. White Cloud, Thompson Creek, Cannivan, and Big Ben lie in a belt that parallels the Colorado Belt but to the north. Wilson Arm and BC Moly lie on an east-west belt that includes a number of smaller deposits. These belts lie within the general northwest orientation of the area of deposits.

ASSOCIATED PLUTONIC ROCKS

The geologic history of most stockwork molybdenite deposits is not simply a case of igneous intrusion followed by hydrothermal mineralization. Rather, it is most commonly a complex sequence of alternating interspersed stages of igneous intrusion and hydrothermal activity. Intrusion with coeval contact metamorphism and metasomatism of the intruded rock initiates the process. Igneous activity can include the pre-, intra-, and post-ore intrusions, forming a complex stock, since successive igneous injections each could be followed or accompanied by a hydrothermal phase (as noted by Wallace, 1974).

Petrography: The intrusions are generally slightly more acidic than those related to porphyry copper deposits; some intrusions of granitic composition may be present although composition could include alaskite, granite, quartz monzonite, granodiorite, or quartz diorite (King, et al., 1973). The phase closest to ore is usually a porphyritic quartz-rich potassic pluton; if the pluton is a complex consisting of several phases, the phases closest to ore spatially and temporally are most often potassic and siliceous. The evolution of a magmatic sequence predictable by differentiation with molybdenum accompanying the end members raises the question of the importance of this mechanism in molybdenum concentration.

Porphyritic textures are predominant with the matrix varying from aphanitic to fine-grained phaneritic. With alteration of a large percentage of phenocrysts and a phaneritic matrix the porphyritic texture can sometimes be recognized only with microscopic examination. Silicate hydrothermal alteration products may also synthesize a porphyritic texture. Quartz invariably is included with the phenocrysts, as is orthoclase or microcline, which occurs in an important minority of deposits.

Morphology: The stocks vary from circular to elliptical at surface with diameters generally about 1000±500 m. Larger intrusions do occur, however. At Questa (Schilling, 1956) a north-south intrusive 3000 m long has a westward bulge that extends below the surface for more than 1500 m. The molybdenite ore body is associated with this western bulge. In other places (Woodcock, et al., 1966) stocks with as little diameter as 250 m can have associated stockwork molybdenite mineralization.

The pluton at Urad is an interesting variation. A postore stock approximately 300 m in diameter cuts off the Urad ore body. This stock enlarges considerably below the surface and becomes the host for the deep Henderson ore body (Wallace, 1974). In other camps (e.g., Endako) only dikes occur (Drummond and Kimura, 1969). Boss Mountain (Soregaroli, 1975) is a breccia pipe penetrating a zoned batholith.

Hydrothermal Stages: The history of a stockwork molybdenite deposit is a complex sequence of interspersed and alternating igneous phases and hydrothermal stages (Fig. 52). Three main stages each divisible into substages may be recognizable. The first includes intrusion of the major part of the stock and the coeval contact metasomatism of the surrounding country rock, which forms adjacent to either barren or productive plutons. The effects of contact metasomatism are most obvious in alumina-iron-rich clastics (shales and graywacke) and in lime-rich sediments. Biotite-plagioclase hornfels are generally formed in the former case; lime-silicate skarns in the latter.

Pyrrhotite, pyrite, or both occur in traces throughout the hornfels and lime silicate minerals. Plagioclase within the hornfels zone has a composition similar to that found within the stock.

The second main stage of alteration may be early deuteric effects that are distinguishable from hydrothermal. Fine-grained sec-

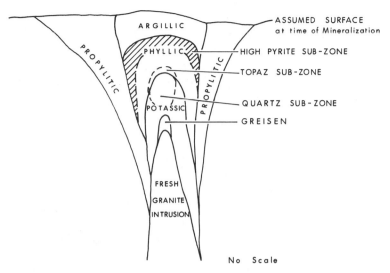

Fig. 52. Generalized section showing alteration zoning in a large Cordilleran porphyry molybdenum deposit. Smaller deposits differ from the larger ones by omission of one or more zones.

ondary biotite may form as a late stage deuteric alteration product during the crystallization history of intrusive rock. Hornblende or pyroxene phenocrysts may be converted to a fine-grained biotite. This can occur anywhere within the intrusive complex, including those places outside the zone of intense hydrothermal alteration where the intrusive rock is relatively unaffected by hydrothermal alteration.

Subsequent to and distinct from formation of the stock with its contact metasomatic aureole and deuteric changes is the main phase of hydrothermal activity, which is a complex event with interspersed alteration, mineralization, and intermineral intrusion. This event forms zones of intense concentric alteration and mineralization. The major minerals, including several sets of quartz, quartz-molybdenite, quartz-pyrite, pyrite, and quartz-pyrite-molybdenite veinlets, are introduced in a number of substages. Additionally some of the late relatively small plutons can contain disseminated molybdenite and fluorite. For example, late alaskite dikes regularly contain disseminated molybdenite rosettes.

Polymetallic sulfide veins form late in the general history and in many cases are separated from the main stage of hydrothermal alteration and mineralization by barren postore intrusions. Polymetallic veins generally have a gangue of quartz (e.g., BC Moly) or of carbonate (e.g., rhodochrosite at Urad). The sulfide minerals include galena, sphalerite, pyrite, chalcopyrite, and —more rarely—a sulfosalt such as tetrahedrite and a bismuth mineral (usually a lead-bismuth sulfosalt). These veins usually occur in a fairly persistent set of fractures compared to the pyrite, quartz, and molybdenite veinlets, which occur in a large number of fracture sets.

Distinct from the polymetallic veins are ribbon-quartz veins consisting of alternating bands of quartz and molybdenite. These represent multiple-stage mineralization controlled by one or more strong fractures. The veins are 5-30 cm thick. Two notable but contrasting examples are worth mentioning. At Endako much molybdenite occurs in a series of subparallel veins 15-100 cm thick. At the Tidewater property near Alice Arm one vein as much as 5 m thick extends about 300 m outward from the stock contact. This single vein was mined during World War I.

Hydrothermal Alteration: The zones of hydrothermal alteration are shown in the idealized section in Fig. 52. Generalizing outward from a fresh intrusive core these are:

(1) Greisen.
(2) Quartz-K-feldspar, including a quartz-topaz subzone.
(3) Phyllic, including a topaz-magnetite subzone.
(4) Argillic.
(5) Propylitic (interpreted in some deposits).

Smaller deposits may not contain all the alteration zones listed.

Central to the alteration assemblages is a fresh stock capped by a postmineral greisen. Ore mineralization overlies the greisen and is associated with the upper limits of a zone characterized by secondary quartz, biotite, K-feldspar, and minor remnants of sericite. This alteration assemblage is similar to the potassic zone of Creasey (1959) and his terminology is used. The quartz:K-feldspar ratio can vary throughout the zone. A plot of either mineral separately shows a very erratic pattern; however, a plot of both minerals combined shows a fairly uniform circular central core. At Climax the central parts of these quartz-biotite-K-feldspar zones include essentially one large mass of silica underlying the ore bodies. In smaller deposits veinlets containing K-feldspar selvages with or without biotite may comprise the central potassic zone. Most major deposits contain a topaz-bearing subzone as part of the upper reaches of the potassic zone. Fluorite may accompany the topaz or replace it.

The phyllic zone contains seritization and silicification of the host rocks. Sericite in matlike fine-grained pervasive intergrowths is characteristic. Pyrite is prominent in this zone and the area of strongest pyrite commonly rims the ore mineralization.

The argillic zone of alteration generally is not conspicuous in hand specimens. Under the microscope, however, the plagioclases exhibit kaolinite, illite, and sericite replacement. Calcite and fluorite may also occur in this zone.

The propylitic zone is characterized by chlorite, epidote, and calcite where it is developed in igneous rocks.

Mineralization: Economic mineralization in porphyry molybdenum deposits includes lithophiles in addition to molybdenum. Tungsten commonly occurs with the molybdenum although molybdenum:tungsten ratios of less than ten are unknown. Wolframite is the most common mineral and may be zoned with huebnerite appearing close to the heat source. Ferberite occurs in the margins of the molybdenum ore bodies.

Tin is present in some deposits but a molybdenum:tin ratio of less than 50 has not been observed and tin has only been recovered from Climax.

Rhenium in molybdenite from porphyry molybdenum deposits is depressed compared to molybdenite from porphyry copper occurrences. Porphyry molybdenum molybdenite commonly has a molybdenum:rhenium ratio in excess of 500.

Gold and silver occur in trace amounts.

Copper may occur with other base metal sulfides in porphyry molybdenum deposits but rarely is common enough to justify separation as a flotation concentrate. The paucity of copper in disseminated molybdenite deposits is characteristic.

A gangue characterized by quartz and fluorite but including silicates distinguishes porphyry molybdenum from porphyry copper deposits. Both minerals may occur in porphyry copper ore zones but occur more commonly with molybdenum mineralization. Both are also more prominent in ore zones with molybdenite since the quartz:sulfide and fluorite:sulfide ratios are larger.

The gangue and sulfide minerals in porphyry molybdenite deposits fill tectonic tension openings either as breccia columns or as stockworks.

Fluid inclusion studies suggest that the ore minerals and the potassic alteration zone were developed in a highly saline fluid. The phyllic zone adjacent to the potassic appears to have formed from less saline fluids.

SPECIFIC EXAMPLES

A number of distinct morphological types occur within the broad outline of deposit characteristics summarized. Depending on size, these types may demonstrate omission of parts of the intrusion and alteration stages. The three described in this chapter contain characteristics representative of other deposits within the porphyry molybdenum classification.

BC Moly (Lime Creek), British Columbia

Woodcock, et al. (1966) describe a cluster of porphyry stocks with associated stockwork molybdenite that occurs immediately east of the Coast crystalline complex at the head of a British Columbia fiord called Alice Arm (Fig. 53). East of the Arm are three mineralized intrusive centers including Roundy Creek, Lime Creek, and Bell Molybdenum.

West of the arm on the same general east-west trend lies Tidewater Molybdenum. Further west on Wilson Arm lies the Wilson Arm deposit, possibly the largest in this belt. These five mineralized stocks are each 5-50 km apart. About 16 km to the north lies the Ajax property adjacent to the Dak River in an area of different regional geologic setting. The three deposits to the east of the Arm are in Mesozoic clastic sediments that trend east to northeast and include graywackes, argillites, and slates. These three are at the locus of a sharp change northward to strata of the Jurassic Hazelton group. These Hazelton rocks trend north to northwesterly and consist of sedimentary rocks that are similar but have interbedded volcanic members.

Intrusive Center: Molybdenite mineralization at BC Moly (Lime Creek) is associated

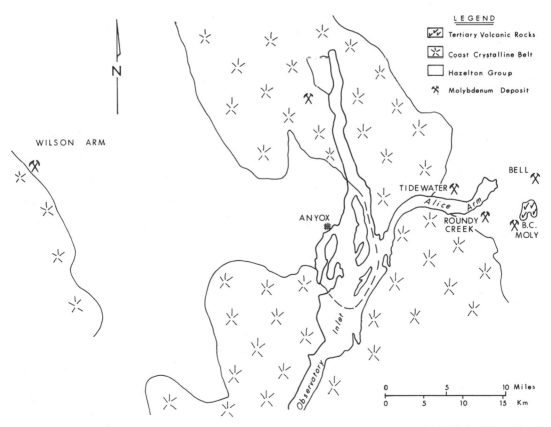

Fig. 53. Location of significant porphyry molybdenum deposits in the vicinity of BC Moly (Lime Creek).

with a small zoned elliptical stock of quartz monzonite-quartz diorite composition that intrudes siltstones and graywackes of Late Jurassic/Early Cretaceous age.

The main stock is composed of granitoid rocks of several types and ages with a central zone of quartz monzonite porphyry. Several types of quartz monzonite porphyry can be distinguished on the basis of texture in the central part of the stock. The rock is essentially medium-grained and leucocratic with euhedral to subhedral phenocrysts of normally zoned plagioclase (An_{25-30}) and poikilitic orthoclase making up the major portion of the rock. Hornblende and biotite are the chief mafic minerals. The main stock, which is 1000 m long and composed largely of porphyritic rocks, has an older eastern appendage about 500 m long composed of quartz diorite with normally zoned plagioclase (An_{42-45}).

Quartz diorite also forms much of the western and southeastern parts of the main stock. This is a medium-grained white-to-gray massive rock with sparse phenocrysts of plagioclase. Fine-grained secondary biotite has replaced the hornblende crystals in much of the rock. In places large orthoclase crystals have formed, many more than one centimeter across. These megacrysts contain relicts of plagioclase and mafic minerals.

Intrusive into all rock types and apparently confined to the northern half of the main stock are irregular lenses and dikes of relatively fined-grained quartz monzonite and granodiorite porphyry and intrusive breccias. These intrusives of intramineral age commonly contain angular fragments of biotite hornfels, quartz monzonite porphyry, quartz diorite, and alaskite in a fine-grained granulated matrix.

Dikes and lenses of white-to-pink equigranular alaskite intrude the quartz monzonite porphyries and the quartz diorite, particularly in the contact areas of the main stock. This rock consists essentially of anhedral quartz and K-feldspar and commonly contains disseminated crystals or rosettes of molybdenite with occasional crystals of fluospar. Disseminated molybdenite mineralization of this type can significantly enhance the grade of the stockwork deposit.

The latest granitic phase is represented by a postmolybdenite quartz-feldspar alaskite porphyry situated beneath the northeast part of the stock. This rock, intersected only by drilling about 200 m below the original surface level, apparently terminates the ore grade mineralization of the northeastern part of the ore zone.

Lamprophyre dikes varying in width from 0.6-9 m (2-30 ft) cut all rocks in the main stock but are especially abundant near the eastern contact. These dikes occur in northeasterly-trending swarms, include both biotite and pyroxene varieties, and have sharp chilled contacts.

Alteration: The siltstones and graywackes in the general region contain premineral chlorite, sericite, minor epidote, and albite plagioclase, placing them in the green schists metamorphic facies. Emplacement of the stock was accompanied by contact metasomatism of the graywackes to biotite hornfels, which contain up to 30% biotite near the stock contact. Outward the biotite content drops off to zero at the biotite line 500-1000 m away from the stock. Adjacent to the stock subsequent hydrothermal alteration has converted some biotite to sericite. A similar zone of biotite hornfels occurs adjacent to the Coast Range batholith.

Hydrothermal alteration is represented largely by concentrically arranged secondary zones of quartz, orthoclase, and sericite. These minerals form an almost circular zone of intense alteration centered in the northern part of the Lime Creek stock. Hydrothermal alteration trends toward a complete replacement within the central part of the zone by quartz and orthoclase in varying proportions both as veinlets and as pervasive alteration. Any plagioclase remnants not converted to orthoclase within this are completely sericitized (Fig. 54).

Secondary orthoclase rims mineralized quartz veinlets and occurs as grains (as much

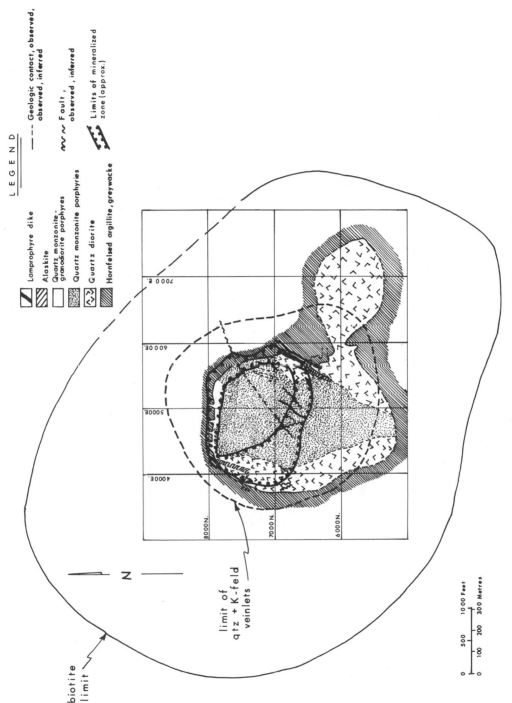

Fig. 54. Generalized geologic map of the BC Moly deposit.

as 5 m across) replacing plagioclase in the rock matrix.

The central intense potassic zone changes quite abruptly to an outer zone of less intense alteration that includes sericitization of plagioclase plus abundant quartz-orthoclase veinlets. The outer limit of the quartz-orthoclase veinlets forms a circular boundary with a diameter of approximately 1000 m. The extent of sericite alteration decreases outward from the orthoclase zone in the stock and in the southern part of the stock only minor sericite and clay alteration are present. Within the biotite hornfels sericite occurs primarily along small quartz veinlets and small fractures.

Argillic and sericite alteration of plagioclase feldspar is relatively intense in and adjacent to northeast-striking faults and shears within every part of the alteration zone.

Within the alteration zone the texture of the porphyries changes. The matrix is recrystallized to a coarser grain size and the phenocrysts are reduced in size by replacement. Thus there appears to be a trend toward an equigranular rock. This end point is never reached and the resulting rock has an almost seriate texture of highly irregular crystals.

Mineralization: The zone of molybdenite mineralization forms a ring pattern slightly elliptical in outline and elongated in an east-west direction. This ring occurs in the outer parts of and outside the intense quartz-orthoclase alteration zone where it makes transition with the phyllic zone. The annular mineralized zone conforms roughly to the north, east, and west contacts of the stocks while the southern part of the zone cuts across the stock at midpoint. The best mineralization grades in the ring are adjacent to the hornfels contact. Molybdenite content fades out toward the center of the zone with the barren core containing only traces.

Molybdenite mineralization occurs along the boundaries of 0.3-0.6 cm-wide quartz veinlets and in hairline fractures. Disseminated molybdenite is found only in the alaskite. Quartz veinlets are closely spaced

and appear randomly oriented in a stockwork pattern but as a general rule the majority of the veins are vertical and strike north-northeast, a prevalent trend for fractures and veins in all the molybdenite deposits of the area. Four separate but superposed substages of molybdenite mineralization are discernable. The first substage is related to the alaskite dikes and is represented by disseminations and rosettes and by fracture fillings of molybdenite. The second and third substages are represented by quartz-orthoclase-pyrite-molybdenite veinlets in a closely spaced stockwork pattern in the northern parts of the stock and adjacent biotite hornfels. Subsequently, quartz monzonite breccias were intruded and these are in turn cut by banded quartz-molybdenite veins as much as 0.3 m (1 ft) thick.

Higher grades of molybdenite mineralization occur in areas of intense fracturing and faulting particularly in the northeast contact area of the stock. However, fracturing has also provided channels for the lamprophyre dike swarms.

The final stage of mineralization is represented by polymetallic sulfide-quartz veins as much as 1 m wide. These veins occur in two conjugate fracture sets that cut the molybdenite zone. The north-northeast set is generally predominant. Less commonly, the northwest set is predominant and in some places both sets are present. The quartz veins contain pyrite, galena, sphalerite, molybdenite, tetrahedrite, chalcopyrite, fluorite, ankeritic dolomite, and a variety of lead bismuth sulfosalts.

Pyrite occurs as disseminations near and within the stock. Pyrite in fractures has been deposited with all substages and can occur in quartz veins, in quartz-molybdenite veins, or alone. The areas of greatest pyrite content form a ring or halo external to but partially overlapping the molybdenite ring.

Deep holes drilled within the stock have encountered anhydrite. This work also indicated a decrease in hydrothermal alteration at depth.

The Urad-Henderson Molybdenite Deposits

The Urad and Henderson molybdenite deposits occur on Red Mountain within the Colorado Mineral Belt, which is a zone 25-100 km wide containing numerous quartz porphyry stocks of Laramide and Tertiary ages. This belt trends northeast across the Rocky Mountains of Colorado and the Urad-Henderson deposits are 50 km from Climax. This description follows Wallace (1974).

General Geology: The Urad-Henderson and Climax ore bodies are related to intrusive centers of mid-Tertiary age (about 30 m.y. at Red Mountain). At these centers composite stocks of rhyolite porphyries intrude Precambrian rocks including various phases of the Silver Plume granite and schists and gneisses of the Idaho Springs formation. Major regional faults occurring near both deposits have had intermittent movement dating back to the Precambrian period. The Mosquito fault at Climax displaces part of the ore deposit.

The Urad-Henderson molybdenite deposits are related to the same intrusive center. However, the Urad deposit crops out at the surface whereas the top of the Henderson deposit lies about 900 m below the surface.

The surface geology can best be visualized by considering it as a sequence of several igneous phases. Three main porphyry units (the East Knob porphyry, the Tungsten Slide igneous complex, and the Square quartz porphyry) of mid-Tertiary age were emplaced prior to formation of the Urad ore deposit. A north-south section across the ore body is shown in Fig. 55. The main fissure for the ore body is thought to be a cone-shaped fracture related to the intrusive center that served as a major distributory for the ore fluids.

The Urad ore body was truncated and partially destroyed by emplacement of the Red Mountain stock; part of it also has been removed by erosion. Although probably only a small percentage of the Urad ore body was destroyed by intrusion of the Red Mountain porphyry, the ore body was only part of a far more extensive but lower grade zone of molybdenite mineralization that formerly existed at Red Mountain. Remnants of this mineralized zone found at various points around the edge of the porphyry column indicate that the zone was pipelike or cylindrical in overall aspect. The ore body proper, although definitely a stockwork, was localized within this low-grade mineralized zone by the influence of the Main Fissure, which also extends into the Red Mountain porphyry stock indicating some postore movement.

Fig. 55 also outlines the Red Mountain porphyry. This elliptical post-Urad plug, 450 x 350 m at the surface, extends downward and merges with a much larger stock, the Urad porphyry.

The most conspicuous rock unit in the Urad area is Red Mountain porphyry. The peak and much of the ridge line of Red Mountain itself are underlain by it. Oxidation of the hydrothermal rhodochrosite contained within has coated much of the surface of this rock black and it contrasts sharply with the limonite-oxidation colors of the other rocks.

The Red Mountain porphyry includes a fine-grained border phase and a coarser-grained central phase, which appears to be the youngest, with relatively sharp contacts between the two. The fine-grained phase contains phenocrysts of quartz, K-feldspar, plagioclase, and some biotite with most of the phenocryst altered and the quartz partially resorbed. Many of the phenocrysts are fractured or fragmented and the matrix is fine-grained to aphanitic. The coarser-grained phase is mineralogically similar but the phenocrysts are less fractured.

The Urad porphyry forms the host rock for much of the Henderson ore body. Essentially all of it has been affected to some degree by hydrothermal alteration and mineralization, so the original texture and composition cannot be accurately determined. However, it does resemble many other porphyries found in the vicinity.

Deep within the Urad porphyry stock lies the Primos porphyry. This aplite porphyry is characterized by the texture of its matrix and

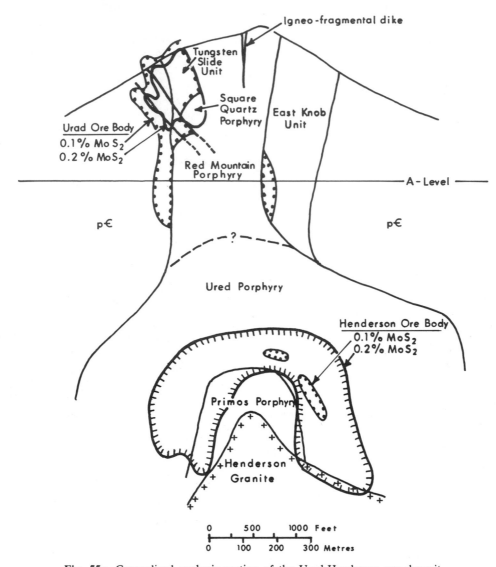

Fig. 55a. Generalized geologic section of the Urad-Henderson ore deposit.

contains phenocrysts of feldspar and quartz.

The Henderson granite forms an inner core for the overlying masses of Primos and Urad porphyries. The various relationships between these three porphyry bodies have not been accurately established because of widespread hydrothermal alteration.

In addition to the main intrusive igneous bodies outlined there are a variety of dikes and small intrusions composed of rhyolite porphyries and intrusive breccias. Some are

exposed at the surface as ring dikes and radiating dikes cutting the Red Mountain porphyry and the older rocks around Red Mountain.

The Ore Bodies: The Urad ore body contained approximately 11.8 million mt (13 million tons) averaging 0.38% MoS_2, with a 0.20% MoS_2 cutoff. The ore zone is arcuate with both ends of the arc terminating against the post-Urad molybdenite Red Mountain porphyry. Although the arcuate

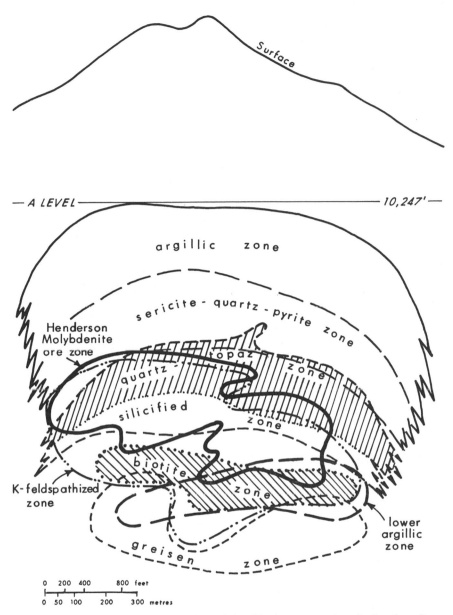

Fig. 55b. Generalized geologic section of the Henderson ore deposit showing alteration zoning.

shape is quite regular, based on the 0.2% cutoff, the higher grade portions of the ore body are erratic because of control, especially in the lower parts of the ore body, by the main fissures.

The highest grades in the Urad ore body occur in the hanging walls of some fault structures. Much of the molybdenite was deposited alone as coating on the walls of tight fractures, in some places forming compound veins as much as 10 cm wide. Some of the molybdenite is accompanied by quartz, pyrite, fluorite, and sericite. However, the general Urad occurrence contrasts sharply with that at Climax where most molybdenite is finely intergrown within quartz veinlets of the stockwork.

The drill-indicated tonnage of the Henderson ore body is in excess of 272 million mt (300 million tons) grading about 0.49% MoS_2. The generalized section in Fig. 55b shows most of the Henderson ore body draped in a general way over the Primos porphyry but it also occurs within the outer and upper parts of the Primos porphyry and partially within the underlying Henderson granite.

Molybdenite within the Henderson ore body occurs in a stockwork with much of it as pure molybdenite or molybdenite in various combinations with quartz, pyrite, and fluorite. In addition some molybdenite is disseminated in the Henderson granite. Apart from this mode of occurrence, some late-stage molybdenite occurs in the quartz-fluorite-pyrite veins that are bordered by greisen alteration within the lower part of the ore body.

Hydrothermal Alteration and Mineralization: MacKenzie (1968) has divided the alteration into two stages including an early pervasive stage associated with the main episode of mineralization. This major alteration stage is separated into several zones shown in the accompanying cross section (Fig. 55).

A K-feldspathic zone is partially co-extensive with the ore body and includes a biotite subzone. Although some late-stage greisen argillic and alteration has subsequently changed much of the lower part of the K-feldspar zone this zone grades downward to essentially fresh rock. The K-feldspathic zone varies in intensity from partial replacement of plagioclase by orthoclase to strong alteration in which K-feldspar and quartz form most of the rock. In the most intensely altered rock primary quartz may be replaced by K-feldspar that includes both orthoclase and microcline.

Most of the primary biotite has been destroyed within the alteration zones. However, some hydrothermal biotite as flakes, replacement patches within feldspars, or veinlets

does occur in the biotite subzone of the K-feldspathic zone.

Superimposed upon and overlying the upper part of the K-feldspathic zone is the silicified zone grading upward into the quartz-topaz zone. The silicified zone generally lies within the lower part of the ore zone rather than completely beneath it as at Climax. The quartz-topaz rock consists of essentially topaz and quartz with varying amounts of sericite and minor pyrite. The silicified zone merges downward into K-feldspar rock.

The quartz-topaz zone grades upward into sericite-quartz-pyrite alteration. This highly altered rock consists primarily of quartz and sericite with pyrite forming 5-15% and with a few relict quartz phenocrysts.

The sericite-quartz-pyrite zone grades upward into the argillic zone. Kaolinite, montmorillonite, and also sericite occur throughout the argillic zone and selectively replace the plagioclase. K-feldspars are relatively fresh or weakly altered by rhodochrosite accompanied by either sericite or kaolinite.

The argillic zone grades upward and outward into a propylitic zone that encompasses a large part of the upper part of Red Mountain, including any possible propylitic zone related to the alteration around the Urad deposit. At surface, propylitic alteration forms an elliptical zone more than 3 km long centered over the Henderson ore body. Within this zone chlorite, epidote, calcite, and pyrite are present within the Silver Plume granite.

Underlying the ore body and partially co-extensive with the lower part of the K-feldspar zone is a lower argillic zone underlain in turn by a greisen zone. However the greisen alteration is not pervasive but occurs as distinct borders on late-stage veins of coarse molybdenite, quartz, and fluorite. The greisen includes quartz, topaz, muscovite, disseminated magnetite, and pyrite. In places fluorite, biotite, and rhodochrosite also occur in the greisen zone, which merges outward from the veins into fresh rock, K-feldspathic rock, or weakly argillized rock of the lower argillic zone. This lower argillic zone is genetically related to the late stage greisen zone and is superimposed on the lower parts of the K-feldspar-biotite zones of the main alteration stage. This late argillic alteration is pervasive but not intense.

Alteration at the Urad deposit is intense. However, the various zones are not as well developed and regular as those associated with the Henderson ore body and overlapping and mixing of alteration types is common. The alteration zoning is similar to that at the Henderson deposit including the central K-feldspar zone followed successively outward by a silicified zone, a quartz-topaz zone, a sericite-quartz-pyrite zone, and an argillic zone. The K-feldspar alteration occurs both in the Tertiary porphyries and the Silver Plume granite and is generally central to the highest grade ore. Within the Silver Plume granite microcline is converted to large irregularly shaped patches of orthoclase.

At the Urad deposit the quartz-rhodochrosite-fluorite with the associated sulfides, pyrite, sphalerite, and galena form a late-stage alteration and replace all rocks including the postore Red Mountain porphyry.

Pyrite forms a definite zone that lies above but also partially overlaps the upper limits of the Henderson ore body. This zone is as much as 600 m thick with abundant pyrite (concentrations of 10% or higher) occurring in the lower 300 m of the zone. Some of the pyrite occurs with molybdenite in the veins and was probably contemporaneous to it; however most appears to occur in younger crosscutting veins.

Magnetite is associated with the Henderson ore body and is concentrated in two specific zones, an upper zone of massive magnetite-pyrite veins and a lower zone in which magnetite occurs as disseminations in the greisen-altered wall rocks of late veins. The upper zone is about 60 m thick and overlaps the lower portion of the pyrite zone and the upper portion of the molybdenite zone (Fig. 55). Weaker magnetite mineralization extends locally downward for several hundred meters into the molybdenite zone.

Although a prominent characteristic of the Henderson deposit, magnetite is not commonly associated with the Urad deposit.

Fluorite is ubiquitous in and near the Tertiary rocks of the Urad deposit and present in all rocks of the Henderson deposit. It occurs in veins, generally as cubic crystals and as disseminated grains or replacements within the Tertiary porphyries.

Topaz occurs in veins or replacements in all the mineralized rocks of the Henderson deposit and is commonly associated with pyrite in the pyrite veins. General distribution seems to be between the strongest pyrite mineralization and the strongest molybdenite mineralization. In the quartz-topaz alteration zone of the Henderson ore deposit topaz can form as much as a third of the rock. Similar completely altered rocks also occur above the Urad deposit.

Wolframite forms veinlike coatings on fracture faces of highly silicified Tertiary rocks above and west of the Urad ore body, generally in the area of highest pyrite content. Ferberite is not a common mineral near the Henderson deposit; on the other hand huebnerite appears erratically, largely concentrated within the molybdenite zone. The huebnerite is largely restricted to the pyrite or topaz veins that cut and offset the molybdenite veins.

Late polymetallic mineralization, galena and sphalerite in a rhodochrosite gangue, is associated with both the Urad and the Henderson deposits. These minerals in combination with quartz, pyrite, and fluorite occur as veins and as replacements in the ore-bearing rocks and the Red Mountain porphyry at the Urad deposit. Potash feldspar is the mineral generally replaced. At the Henderson ore body this late-stage mineralization generally takes the form of veins.

Boss Mountain Molybdenum Deposit, British Columbia

The Boss Mountain molybdenum deposit exemplifies molybdenite mineralization and related alteration with direct structural control by a breccia pipe and related fractures. This brief summary is largely from Soregaroli (1975).

General Geology: Boss Mountain lies within the eastern part of the composite

Takomkane batholith, a zoned massif of diorite, quartz diorite, and granodiorite that crops out over an area of about 1500 sq km and intrudes volcanic rocks of Triassic age. Andesite dikes as much as 3 m (10 ft) wide striking generally northwesterly cut the batholith. The granodiorite has been dated at 187 m.y., establishing an Upper Triassic age for the batholith.

Intruding the eastern part of the Takomkane batholith is the Boss Mountain stock, an elliptical pluton of quartz-monzonite porphyry approximately 1220 x 610 m (4000 x 2000 ft). This quartz-monzonite porphyry is of medium to coarse grain with phenocrysts of quartz, plagioclase, and orthoclase. Stock contacts are sharp with a chill zone 2-3 m wide at the margin. Dikes of rhyolite porphyry and rhyolite, which are apothyses of the Boss Mountain stock, occur near the molybdenite deposits.

Three phases of breccia are recognized in the Boss Mountain deposit and referred to as the Boss breccias. Phase I breccia is found in two irregular bodies in and near the molybdenite deposits. Angular or subrounded fragments generally 1-4 cm across of granodiorite and porphyritic biotite granodiorite are set in a gray-to-black comminuted matrix of the same composition. Some fragments of rhyolite porphyry, andesite porphyry, garnet-hornblende veinlets, and sugary quartz veins are also present. The Phase I breccia cuts rhyolite porphyry dikes and fragments of these dikes are common within the breccia. In some places this breccia has a rhyolite porphyry matrix, suggesting that intrusions of rhyolite porphyry dikes and formation of Phase I breccia are cogenetic. Some non-porphyritic rhyolite dikes cut Phase I breccia and occur as fragments in later breccia phases indicating that rhyolite dike emplacement commenced before and continued during and after formation of the Phase I breccia.

The Phase II or quartz breccia occurs in well defined steeply plunging lenticular or pipelike masses that have been traced from surface downward to termination at the Boss Mountain stock, a vertical distance exceeding 366 m (1200 ft). The fragments, generally 2-5 cm across, are commonly granodiorite but can also include all prior rock types. These are set in a matrix of quartz and sulfide. Some contacts of the quartz breccia body are transitional and grade outward to quartz stockwork with nonrotated fragments and then into massive rock with widely separated quartz veins.

Phase III breccia is composed of angular to well-rounded fragments in a gray to light brown siliceous matrix that probably was derived from comminution of quartz breccia. This matrix generally is somewhat porous and commonly contains scattered crystals of orthoclase, biotite, stibnite, and disseminated molybdenite.

The hydrothermal biotite developed during Phase III breccia formation has been dated at 105 ± 2 m.y., establishing a mid-Cretaceous age for the ore deposit.

Alteration and Mineralization: Soregaroli (1975) recognized eight distinct periods of fracturing. Group 1 fractures cut rocks of batholith and were filled with quartz and some pyrite. Group 2 fractures are associated with Phase II (quartz) breccia, containing quartz with minor pyrite and orthoclase. Group 3 fractures are associated with Phase III breccia and are filled with quartz-molybdenite veins. Group 4 fractures include early-stage quartz veins with molybdenite, orthoclase, and pyrite and somewhat later-stage quartz veins with orthoclase, pyrite, bismuth, tungsten, and copper minerals. Group 5 fractures are filled with ribbonlike quartz-molybdenite veins that comprise several ore bodies. Group 6 fractures are accompanied by intense chloritization and the wall rock adjacent to some fractures is altered to clay minerals. Groups 7 and 8 fractures are post-mineral in age and contain abundant open spaces, many of which are lined with calcite and zeolites.

Six distinct stages of rock alteration are recognized in the Boss Mountain deposits, some genetically related to molybdenite mineralization. These include:
1) Garnet-hornblende.
2) Biotite.
3) Quartz-sericite-pyrite-orthoclase-chlorite.

4) Chlorite-talc.
5) Epidote-chlorite.
6) Zeolite-calcite-clay.

The initial stage of alteration related to the Boss Mountain deposits is a garnet-hornblende assemblage that occurs as narrow veinlets that have replaced mylonite zones developed within the granodiorite and andesite dikes around the Boss Mountain stock. Each veinlet consists of a core of reddish brown andradite bordered by a narrow black selvage of hornblende and magnetite. Epidote forms a major portion of some veinlets but is lacking in most.

Secondary biotite occurs in a large zone around the Boss Mountain stock. Biotite metasomatism is most intense in and near the breccias, decreasing with distance from them. Biotite development east of the Boss Mountain stock is especially weak and can only be detected in thin sections. Green biotite occurs as mosaic ovoids, irregular patches, and veinlets in altered andesite dikes. Pale brown biotite occurs as single crystals in the matrix of the Phase III breccia and as aggregates replacing andesite fragments within this breccia. Much of the biotite is a replacement of magmatic and hydrothermal hornblende and magmatic biotite.

Quartz-sericite-pyrite-orthoclase-chlorite is restricted in distribution and varies considerably in intensity. Group 4 quartz veins with this type of alteration occur throughout the mine area, where the alteration has been superimposed on biotite-altered rocks. However, the most intense and widespread area of this alteration is within the Boss Mountain stock.

Chlorite-talc alteration is restricted to dike and veinlike bodies that replace comminuted rock matrix along Group 6 fractures. Intensely biotized andesite dikes were particularly susceptible to this alteration.

Widespread epidote-chlorite alteration occurs sporadically within the limits defined by pyrite distribution, but large areas within the pyrite zone do not contain this alteration. The epidote occurs as irregular discontinuous veinlets adjacent to which the rock is chloritized for a few inches.

Stage 6 alteration (zeolite-calcite-clay) presumably is related to emplacement of alkali basalt dikes. Postmineral fractures of the mine area contain fillings and encrustations of zeolite and calcite and masses of stilbite replace plagioclase in the matrix and fragments of the Phase III breccia. Clay minerals developed in plagioclase adjacent to these zeolite- and calcite-filled fractures.

Most of the molybdenite ore in the Boss Mountain deposit is confined to breccia bodies or vein systems. The main breccia zone is a crudely lenticular vertical body that strikes northeast and is composed of quartz breccia and Phase III breccia. From the surface the ore zone separates downward into two roots, one of which spans the entire vertical extent of the breccia, a distance exceeding 400 m. This ore zone is about 200 m long at the surface but has a lesser length at lower levels. A fracture zone forms a bulbous hood that envelops part of the northwest quartz breccia. Contacts with the underlying quartz breccia and overlying granodiorite are gradational to stockworks.

In places narrow veins within the mine area occur as subparallel swarms, referred to as stringer zones, which constitute low-grade ore bodies. Such ore occurs as a gently dipping arcuate skirt around the north and west margins of the breccia zone. Other less important breccia and stringer zones are also present.

In addition one very high-grade quartz-molybdenite vein occurs predominately within an altered and sheared andesite dike. The width ranges from one to more than three meters. This intensely altered dike is composed of chlorite, talc, and carbonate, and molybdenite is distributed erratically across the width of the dike with local averages of several percent.

CONCLUSIONS

Porphyry molybdenum and porphyry copper deposits have much in common. Each is associated with a heat source most commonly identified as a stock. Mineralization in each occurs in tension fractures and in each the alteration zone has an alkali-rich core zone. Potassium feldspar and biotite are common secondary silicates developed in the alteration zone.

Metallization in porphyry molybdenum deposits is distinct, on the other hand. The economic minerals rarely include copper but do include lithophiles other than molybdenum. Tungsten and tin are both minor constituents in many disseminated molybdenum deposits. Rhenium, on the other hand, is consistently higher in porphyry copper molyb-

denite than in this mineral in porphyry molybdenum deposits.

Gangue minerals and alteration zones in porphyry molybdenum may contain significant topaz and fluorite. Both minerals tend to be insignificant in most porphyry copper deposits. Quartz is much more prominent in porphyry molybdenum than in porphyry copper deposits. The differences in metallization and alteration products are sufficient to distinguish the two types.

REFERENCES AND BIBLIOGRAPHY

Anon., 1974, "Cyprus Exploration," *Engineering and Mining Journal*, July, p. 27.

Anon., 1976, "Wilson Arm," *Wall Street Journal*, Feb. 7.

Armstrong, R. L., et al., "The White Cloud-Cannivan Porphyry Molybdenum Belt," *Economic Geology*, Vol. 72, in press.

Bennett, E. H., 1973, "Petroiogy and Trace Element Distribution of the White Cloud Stock," Ph.D. Thesis, University of Idaho, Moscow, ID, 172 pp.

Clark, K. F., 1972, "Stockwork Molybdenum Deposits in the Western Cordillera of North America," *Economic Geology*, Vol. 67, p. 731.

Creasey, S. C., and Scholz, E. A., 1945, "Big Ben Molybdenum Deposits, Neihart County, Montana," Open File Report, US Geological Survey, p. 29.

Drummond, A. D., and Kimura, E. T., 1969, "Hydrothermal Alteration at Endako," Bulletin No. 6, Canadian Institute of Mining and Metallurgy, p. 1.

Giles, D. L., 1975, "Geology and Isotope Geochemistry of the Lime Creek Ore Body, British Columbia," SME-AIME, Preprint No. 75S72, AIME Annual Meeting, New York, p. 34.

Hall, W. E., Friedman, I., and Nash, J. T., 1974, "Fluid Inclusion and Light Stable Isotope Study of the Climax Molybdenum Deposits," *Economic Geology*, Vol. 69, p. 894.

King, R. U., 1970, "Molybdenum in the United States," Miscellaneous Field Report No. MR-55, US Geological Survey.

King, R. U., Shane, D. R., and MacKevett, E. M., 1973, "Molybdenum," *US Mineral Resources*, Professional Paper No. 820, US Geological Survey.

Kirkham, R. V., 1969, "A Mineralogical and Geochemical Study of the Zonal Distribution of

Ores in the Hudson Bay Range, British Columbia," Ph.D. Thesis, University of Wisconsin, Madison, WI, p. 210.

Kruschov, N. A., 1959, "Classification of Molybdenum Deposits," *Geology of Ore Deposits*, Academy of Science, Moscow, USSR, No. 6, pp. 52-67.

MacKenzie, W. B., 1968, "Red Mountain Molybdenum Deposits," Ph.D. Thesis.

Ross, C. P., 1937, "Geology of the Bayhorse Region, Idaho," Bulletin No. 877, US Geological Survey, p. 161.

Schilling, J. H., 1956, "Geology of the Questa Molybdenum Mine Area," Bulletin No. 51, New Mexico Bureau of Mines, p. 85.

Soregaroli, A. E., 1975, "Geology and Genesis of the Boss Mountain Molybdenum Deposit in the Southern Rocky Mountain Molybdenum Province," *Abstracts*, Geological Society of America, Vol. 7, p. 646.

"United Nations Report of Exploration in Sonora, Mexico," 1970, United Nations, New York, NY, p. 96.

Wallace, S. R., 1974, "The Henderson Ore Body— Elements of Discovery, Reflection," *Trans. SME/AIME*, Vol. 256, p. 216.

White, W. H., 1976, "The Adanac Deposit," *Porphyry Deposits of the Canadian Cordillera*, Special Vol. 15, Canadian Institute of Mining and Metallurgy.

Woodstock, J. R., Carter, N. C., and Ney, C. S., 1966, "Molybdenum Deposits at Alice Arms," *Tectonic History and Mineral Deposits of the Western Cordillera*, Special Vol. 8, Canadian Institute of Mining and Metallurgy, pp. 335-339.

Worthington, J., 1977, "Molybdenum Mineralization at Cannivan Gulch, Montana," *Abstracts*, Geological Association of Canada, Vol. 2, p. 172.

8

Discussion and Conclusions Regarding Porphyry Copper Deposits of the Western Hemisphere

CONTENTS

INTRODUCTION

Data available on prophyry copper occurrences in the Andean, Caribbean, Appalachian, and Cordilleran orogens may be categorized in a manner to suggest a hypothetical evolution for deposits that occur in the western hemisphere. A classification of deposits is also suggested.

A classification dependent on internal structure of a porphyry copper deposit would lead to two broad categories, breccia pipe and stockwork models. A classification of this type is examined herein but it is not universally accepted in the literature.

Modeling based on igneous petrography as well as metasomatism attendant with the sulfide mineralization is more widely accepted by economic geologists whose evaluation of porphyry systems is influenced by the presence or absence of a phyllic zone. Lowell and Guilbert model deposits ordinarily include a phyllic zone, whereas those of the

diorite model do not. Position of the phyllic zone in the alteration sequence is a significant ore guide. Should the phyllic zone be absent, dependence on alteration zoning to identify metallization is weakened. Presence or absence of the phyllic zone therefore is a significant parameter in the economic evaluation of a porphyry copper deposit (Hollister, et al., 1975).

The Lowell and Guilbert (1970) model commonly includes a phyllic zone in quartz-bearing rocks. Parameters of both calc-alkalic and the quartz monzonite models mentioned elsewhere in this volume are sufficiently close to the Lowell and Guilbert (1970) model that it should be expanded to encompass them. The diorite model remains separate and distinct because deposits of this type are characterized by an absence of quartz in the pluton, commonly highly gold: copper ratios, low molybdenum content, and regular absence of sericite in the alteration zones. Therefore, for the benefit of those economic geologists who use alteration zoning in evaluation, the Lowell and Guilbert and the diorite models are defined. Some deposits of the Lowell and Guilbert model (e.g., Safford and Ajo in Arizona) may lack a substantial phyllic zone and some deposits may incorporate igneous rocks from both models (e.g., Esperanza-Sierrita or Ray in Arizona). The diorite model is not widely used because of confusion as to its application. It may be employed effectively as a prospecting tool in some terranes, however, and is therefore mentioned.

Synthesis of this section has greatly benefited from discussions with A. W. Rose and A. J. Sinclair.

STRUCTURAL CLASSIFICATION

Internal structure may be the basis for classifying porphyry copper deposits into breccia pipe or stockwork types. Categorizing deposits after the type of dominant structure is not always possible, however. Structure is not always mapped in individual deposits. In some the presence of both stockworks and breccias inhibits a simple classification. Nevertheless Sillitoe and Sawkins (1971), Sutherland Brown (1969), and Hollister (1974) note that the stockwork and breccia pipe categories are locally meaningful in the western hemisphere.

Breccia Pipes

Breccia pipes present in deposits with a dominant stockwork structure are generally small and play a subordinate roll in the localization of metal sulfide. Large breccia pipes, on the other hand, may have stockwork structures ringing their periphery and metal sulfide is primarily controlled by the pipe. The distinction between large and small pipes is economically significant.

Small Breccia Pipes: Small breccia pipes accompany stockwork deposits and may have configurations and other details different from those found in large pipes. Small pipes may have distinct bottoms (Mills, 1972), blind tops (Perry, 1961), and grade into breccia dikes along strike (Mitcham, 1974).

The bottom configuration of small pipes described by Locke (1926), Joralemon (1952), Kuhn (1941), and Kents (1961) suggests the breccia may terminate downward in a funnel-shaped neck or grade into a series of tight fractures.

Top configuration of small breccia pipes may show that the pipe itself is *blind*, having developed without relief to the present surface (Perry, 1961). Breccia pipes penetrating layered rocks may not be accompanied by arching. Small breccia pipes that do reach the surface may have significantly enlarged cross sections at depth and these features lead Mills (1972) to the conclusion that hypotheses advocating origin by explosion or withdrawal of underlying magma cannot be universally applied to small pipes.

Small breccia pipes (e.g., those at Highland Valley, Esperanza, Cariboo Bell, and Chaucha) tend to be monolithologic with the fragments largely or entirely of one rock type. The lack of mixing of fragments suggests a restricted source for the rock fragments, limited movement of the fragments, and a limited vertical extent for the pipe itself. Because only one rock type commonly is involved no evidence for movement up or down can be seen. Yet examples are known that show downward movement of fragments (Mills, 1972).

Fragments in small breccia pipes tend to be angular. Some fines may be present, although close inspection of the matrix shows the pipe to be deficient in the size fractions of sand and smaller gains (Bryner, 1961).

Generally the smaller the pipe the greater its tendency to be elongate or grade into a breccia dike (Bryner, 1961).

Smaller breccia pipes usually are elongate parallel to the strongest veinlet trend in the stockworks of a porphyry system (e.g., Bethlehem).

Large Breccia Pipes: Botton configuration of large breccia pipes remains unknown because no large pipes (e.g., Toquepala, Braden, Casino, or La Caridad) have ever been bottomed. Nor have the roofs of any of these pipes been seen. Large pipes tend to have fragments of more than one type of rock and mixing is a characteristic. Mixing may have taken place at Stikine but extreme potassic metasomatism appears to have masked this feature. Perry (1961) presents evidence that magma withdrawal may be one possible cause of the breccia formation with rock fragments collapsing into the void left by the withdrawal. Richard and Courtright (1958) leave open the possibility of injection of the breccia at some point as an explanation for mixing of fragments. Refragmentation and multistage brecciation with mixing of rock types in fragments appear to be typical of the larger pipes.

Major breccia pipes commonly are ringed with a peripheral annular fracture set, whereas such fractures are absent in the walls of small pipes. A radiating set of tension fractures accompanies the annular fractures. The two sets are compatible with a strong vertical compressive force having acted early in the history of pipe formation. Perry (1961) assumes a magma acted as a diapir to make the initial penetration, whereas Bryner (1961) argues for penetration by stoping. In this view high pressure from thermal fluids was responsible for the piercement. Presence of porphyritic intrusive material in the fragments of most large pipes supports the Perry (1961) hypothesis. It is equally clear, however, that ascending hydrothermal fluids under strong lithostatic pressure may have been responsible for some of the characteristics found in large pipes. The consistent association of tourmaline and pipes suggests that boron is one constituent of the ascending highly saline magmatic-hydrothermal fluids.

Larger pipes also appear to be missing a larger percentage of the fines, assuming they were generated. Because of this Mills (1972) argues that an origin "by solution and collapse" seems the only one compatible with the data. The apparent removal of enormous volumes of rock fines to accommodate rotation of the larger fragments and permit entry of the hydrothermal mineral matrix (and vugs) may be misleading. The fines could have been largely incorporated into matrix silicates as reconstituted gangue minerals rather than completely removed. Some volume increase for the pipe is always assumed

but the fines would clearly be more susceptible than coarse fragments to solution by corrosive hydrothermal fluids. In all pipes quartz is dominant in the matrix and is regularly accompanied by sericite, orthoclase, tourmaline, or other silicates, or some combination thereof. These could have been derived from the fines.

Hydraulic pressure and turbulence from a rising hydrothermal fluid passing through a collapsing column of fragmented rock may also have removed some fines and this mechanism is compatible with all the field evidence for larger breccia pipes. Even though some fines should have reacted with the hydrothermal fluid to form the large quantity of silica and silicates present in the matrix of most pipes, the apparent volume increase is compatible with some overall removal of material. The likelihood that hydraulic pressure from a rising fluid column was involved in breccia pipe formation raises the possibility that a diatreme is a surficial expression of a large breccia pipe.

Most major breccia pipes contain a late-stage upward expanding pebble pipe or pebble dike. Argillic alteration is common in this stage and fragments are distinguished from the breccia pipe (termed ore breccia at Toquepala by Richard and Courtright, 1958) by rounding. Whereas the tourmaline breccia or the earlier sulfide-bearing intramineral ore breccias have angular fragments cemented by a matrix that includes tourmaline and sulfide, the late-stage pebble breccia has rounded fragments cemented by a matrix that includes clay. It may also, however, include minor late-stage tourmaline but usually not sulfide.

Pebble dikes occur in most porphyry deposits. Ordinarily they are unnoticed, whereas pebble pipes with oval or circular cross sections most commonly occur with large tourmaline breccia pipes. Milled and rounded fragments in a mixed fragment body commonly are set in a clay-rich matrix that somewhere includes graded bedding of the fines. Ore minerals may occur in rounded fragments of the pebble pipe (as in the Braden pipe), implying a postmineral origin.

Matrix mineralogy and paragenesis of pebble dikes as a late event suggest that, if they are part of the ore-forming hydrothermal system, they developed in a meteoric-hydrothermal regime. By extrapolation of data obtained elsewhere in oxygen thermometry temperatures of formation of less than 200°C may be inferred for these bodies. Their physical aspect including graded bedding and fragment rounding by autogenous grinding suggests deposition in a liquid (aqueous) invironment.

Stockwork Type Deposits

Stockwork type deposits are prone to association with a porphyry copper that is involved in some way with movement on a major fault. Generally the fault has a dominant strike-slip movement. Numerous examples show the stockwork to be a result of repeated movement on one or more faults and in many of these intrusion of the pluton may be shown to be controlled by the same faults. The stockwork controlled or governed the routing or path of hydrothermal fluids.

In summary stockwork deposits are those porphyries whose ore constituents occur predominantly in vein and veinlet filling. The density of mineralized fractures varies from one deposit to another but the deposits generally have a number of features in common:

1) Each deposit is close to or lies on a major fault.

2) Each deposit has a clearly ascertainable principal set or trend of veinlets.

3) The stockwork fractures were tectonic voids filled with mineral.

4) In each deposit the actual channeling of hydrothermal fluids was controlled by the stockwork fractures and mineral zoning occurs in the stockwork as a function of distance from a heat source.

These may be generalized into several distinct types:

1) Intersecting faults not demonstrably part of a conjugate set as exemplified at Chaucha, Michiquillay, and Highland Valley.

2) Conjugate fractures in either wall of one major fault as shown at Chuquicamata and Copaquire.

3) Shear zone fracturing as demonstrated at Butte.

4) Annular fracturing caused by an intrusion (diapirism) as exemplified by Campana Mahuida, Berg, and Earl.

PORPHYRY COPPER MODELS

Lowell and Güilbert Model

Lowell and Guilbert (1970) presented the first comprehensive model of porphyry copper systems. Their model included porphyry copper deposits associated with calc-alkalic plutons of the granodiorite-quartz monzonite range. Alteration zones in this model could include all or some of the following, progressing outward from the center of the system: a potassic core (orthoclase-biotite), a phyllic zone (quartz-sericite-pyrite), an argillic zone (kaolin-illite-montmorillonite-pyrite), and an outer propylitic zone (epidote-chlorite). Alteration zoning for this model is also characteristic for intruded rock whose chemical and physical characteristics include most fine- and coarse-grained clastics but not for calcareous rocks. Prospecting for and exploration and development of deposits with characteristics matching those of the Lowell and Guilbert (1970) model utilizes the recognition of these alteration zones.

Petrography: Most plutons associated with the Appalachian, Andean, and southern Cordilleran deposits have low $Na_2O:K_2O$ ratios and therefore many plutons in these areas are in the granodiorite-quartz monzonite range, typical of the original Lowell and Guilbert (1970) model.

Kesler, et al. (1975), Cox, et al. (1975), and Hollister, et al. (1975) point out that porphyry deposits with quartz-diorite plutons having high $Na_2O:K_2O$ ratios occur in the continental margin or island arc environment. These generally are porphyries with megascopically visible quartz and plagioclase. Accompanying sulfides commonly have a high gold:copper ratio but, with rare exceptions such as Sapo Alegre (Cox, et al., 1975), a low molybdenum:copper ratio compared to classical or type examples of the Lowell and Guilbert (1970) model. Fig. 36b (page 104) shows petrographic trends of porphyritic intrusives associated with occurrences in the Cascades. These trends are also representative of porphyry copper deposits in the continental margin belt of Alaska. Petrographic data for porphyry copper deposits in the Caribbean island arc (Kesler, et al., 1975) also show that Caribbean and Cascade deposits have a similar petrography. The Lowell and Guilbert (1970) model should be expanded to include quartz diorite-related deposits exemplified by those in the Caribbean, Cascades, and Alaska.

Alteration: A general statement applicable to deposits fitting the Lowell and Guilbert (1970) model is that alteration geometry and mineralogy are a function of a regional structural fabric, heterogeneity of preore rocks, and the effects of sulfur, hydrogen, sodium, and potassium metasomatism on these rocks. Hydrogen ion concentrations in the altered rock increase outward from the potassic through the phyllic zones but decrease again in the propylitic. Hydrogen ion concentrations also increase with time during the mineralizing event. Sulfur concentrations increase from the potassic, peak in the phyllic, decrease again in the argillic, and reach zero in the propylitic. Potassium metasomatism generally is strongest in the potassic zone but weakens from the core outward through a control of the K/H ratio of the altering fluids.

In detail the potassic zone originally proposed for the Lowell and Guilbert (1970) model may consist most commonly of either an orthoclase-biotite, an orthoclase-chlorite, or a biotite-chlorite zone, although in a few deposits all occur. The potassic zone is missing from a minority of deposits. This zone in Appalachian deposits may contain abundant epidote (e.g., Mariner) and may be peripheral to a fresh unaltered core of the mineralized pluton. In the continental margin belt of Alaska, or the Cascade province of the northern Cordilleran orogen, the central potassic zone could consist of secondary biotite or chlorite and hydrothermal plagioclase with little or no potassium feldspar. In both Alaska and the Cascades alteration silicates present in the potassic zone reflect host quartz diorites having high $Na_2O:K_2O$ ratios. The absence or impoverishment of orthoclase in potassic zones of some deposits along the continental margin of Alaska and Washington is similar to the development of biotite-chlorite-rich potassic zones in the Antilles. Cox (1973) reports possible orthoclase deficiency or absence in the central potassic zone of porphyry copper deposits of Puerto Rico. In his description chlorite follows the secondary biotite but both may be peripheral to a core zone containing abundant hornblende and actinolite. The name potassic zone should be retained for simplicity even for those zones deficient in potassic feldspar.

In deposits with a core central to the potassic zone the core may be either a fresh intrusive or a hornblende-rich zone. The potassic zone itself could consist from the core outward of an epidote-bearing stage, a chlorite-bearing stage, and a biotite-bearing stage, although ordinarily only one of the three stages is developed in any one deposit. No deposit has yet been described that contains all three. Orthoclase may or may not be a constituent of any or all of these stages. Magnetite appears instead of pyrite in some deposits. Secondary microcline is considered an acceptable substitute for orthoclase.

The phyllic zone is also the sericite-quartz-pyrite zone although pyrophyllite or other sericite-appearing minerals may importantly occur. Kaolin may also appear in the phyllic zone; phyllic zone mineralogoy is usually complex in detail. Albite, chlorite, and other hydrated silicates have been reported from this zone.

The argillic zone is characterized by illite, kaolin, montmorillonite, and pyrite. Disseminated carbonate is important in some deposits. Chlorite is the most characteristic mineral developed in the propylitic zone although pyrite, calcite, and epidote commonly occur with it. Albite-rich propylitic zones are also known. Orthoclase has also been identified in the propylitic zone.

Pyrite is most commonly developed within and as a halo to the ore body in the Lowell and Guilbert (1970) model. As noted in this volume pyrrhotite may occur with or instead of pyrite in some deposits. In the Cascades pyrite may exist as an outer halo, whereas pervasive pyrrhotite is found closer to minerals of economic interest. Pyrite is less well developed in the potassic zone than in the phyllic. It may constitute as much as 20% of the phyllic zone, which has the greatest development of pyrite, the strongest silicification, and the most prominent sericite.

The larger the porphyry copper occurrence, the more regular and well developed the alteration zoning. The reverse is equally true since smaller deposits are inclined to show smaller, more erratic, and less complete alteration effects.

Tourmaline: Tourmaline as an epigenetic mineral of hydrothermal origin occurs erratically but commonly associated with breccia pipes in the Andean orogen and in two areas of the Cordilleran orogen (Seward Peninsula arc, which includes the Hogatza plutonic belt of Alaska, and the Sonoran porphyry province of Mexico). Other isolated examples have also been cited besides these belts. In each area tourmaline occurs both as a fracture filling with other gangue minerals and a pervasive alteration mineral disseminated in wall rocks. It is most distinctive as a matrix mineral cementing fragments in breccia pipes. In some deposits it constitutes as much as 5% of the ore and in some larger concentrations as much as 45 million mt (50 million tons) of tourmaline have been deposited. Tourmaline is an alteration mineral ordinarily associated with intrusions in the quartz monzonite-quartz diorite compositional range. The porphyry copper deposits containing tourmaline commonly may be defined within the parameters of the Lowell and Guilbert model. Tourmaline-bearing diorite model porphyries (e.g., Gnat Lake, British Columbia) are rare.

Tourmaline may occur, as at Toquepala, in either the propylitic, argillic, phyllic, or potassic zone of the Lowell and Guilbert model. It is pervasive and appears to have a wide range of pressure-temperature stability. In all zones it appears to preferentially replace original ferromagnesian minerals. Depending on the location of the replacement in the zonal alteration sequence the accompanying minerals may include clay, chlorite, sericite, plagioclase, biotite, or orthoclase. The lack of correlation between tourmaline and any one specific alteration zone sequence makes it of little value as an indicator mineral for defining alteration zones or as an ore guide for copper. Its widespread occurrence as a replacement of original magmatic ferromagnesian minerals suggests that boron in the hydrothermal fluids migrated easily into the wall rocks. The preference of tourmaline to occur as cementing fragments in breccia pipes suggests it is initially derived from a magmatic-hydrothermal fluid. However, its presence in every alteration zone in those porphyry deposits where it does occur implies that boron easily joins components of the meteoric-hydrothermal system. Tourmaline-bearing porphyries generally contain an important amount of disseminated pyrite as well. The deposition of tourmaline-quartz-pyrite commonly commences early, preceding economic metallization. Later deposition of this mixture may also be penecontemporaneous with copper sulfides. The occurrence of intimate fine-grained mixtures of all these minerals in a number of deposits suggests the probability that the hydrothermal fluids that may have been instrumental in the formation of the tectonic openings (breccia columns) also introduced touramline and metallic constituents. Deposition of tourmaline may continue after sulfide mineral mineralization in some deposits (e.g., Toquepala).

Mineralization.: Hypogene copper and molybdenum sulfides in the Lowell and Guilbert (1970) model may occur significantly in either the potassic or the phyllic zone or both. Hypogene copper sulfide appears to be most strongly developed at and near the interface between the potassic and the phyllic zones. Hypogene copper only rarely occurs in the argillic zone and where it does it tends to be in association with other sulfides. The larger the porphyry copper occurrence the more likely is the copper to occur in the argillic zone. However, the smaller the deposit the less likely is the copper to occur outside the phyllic zone.

If bornite occurs in the system it is found in the potassic zone, whereas chalcopyrite has a much broader distribution. If copper is only transported into the upper crust by the sulfur-rich magma it should be expected to occur importantly with potassic zone minerals as it does at Bingham. The appearance of potassic zones barren of copper but ringed with copper-bearing phyllic zones (e.g., Taurus and Pyramid, Alaska) suggests the probability that copper could be removed from the potassic zone by highly saline magmatic-hydrothermal fluids. The rarity with which copper occurs outside the phyllic zone also implies a dumping of that element adjacent to the potassic zone if it is indeed transported from the potassic by magmatic-hydrothermal fluids.

The same copper distribution pattern leaves open the question as to how much wall-rock leaching of copper the meteoric-hydrothermal systems accomplish. Hypogene copper sulfides commonly occur as disseminations intergrown with silicates derived from magmatic-hydrothermal fluids in the potassic zone. Copper in the potassic zone therefore may be identified with the magmatic-hydrothermal fluids.

Outside the potassic zone hypogene copper minerals occur, filling fractures that cut alteration silicates developed from meteoric-hydrothermal-dominant fluids. Hypogene copper occurring in these fractures conceivably could have been deposited from magmatic-hydrothermal or meteoric-hydrothermal fluids. Hypogene mineralization in the Lowell and Guilbert (1970) model phyllic zone consists of sulfide, quartz, and sericite. Anhydrite is commonly present. Similar mineralized fractures generally exist in the potassic zone but dissemination of ore sulfide within the rock at a distance from fractures may not occur in the phyllic. Magnetite, if present, generally occurs in the potassic and propylitic zones but not in the phyllic. Fracture filling in the propylitic and argillic zones may be of the sulfide-quartz type but pyrite-, pyrite-chlorite-, or pyrite-calcite-filled fractures are dominant.

Porphyry copper deposits that are essentially wall rock porphyries (Titley, 1972) may impose alteration and mineralization on a host rock with a chemical composition and physical response markedly different from that of a granodiorite or quartz-monzonite intrusion. One example may be limestones or calcareous sediments converted to skarns (e.g., Mission-Pima, Arizona).

Paragenetic studies of hypogene sulfide minerals invariably show pyrite to be early. Its initial introduction is followed by deposition of the base metal sulfides and in some example may occur after them.

Supergene copper sulfides reflect greater mobility and may occur in the argillic zone as well as the potassic and phyllic. The propylitic zone most commonly is low in or barren of both hypogene and supergene copper sulfides.

The Lowell and Guilbert (1970) model should be expanded to include the variations of alteration and mineralization noted in this volume.

Diorite Model

Those chapters in this volume that describe porphyry copper deposits in Alaska and the northern Cordilleran orogen use the diorite as well as the Lowell and Guilbert model. As originally proposed by Hollister (1974) this model encompasses porphyry copper deposits associated with quartz-free plutons that commonly contain only the potassic and propylitic mineral assemblages. Included are alkalic petrologic suites as well as calc-alkalic mineralized diorite intrusions (Fox, 1975; Ney and Hollister, 1976; Soregaroli, 1975; and Sutherland Brown, 1974). The diorite model has affinities to the island arc model (Portacio, 1974)

Porphyry deposits occurring entirely in andesite and diabase are likely to have alteration sequences found in dioritic plutons. For this reason the phyllic zone mineral assemblage has not been commonly reported in porphyry copper deposits developed in this kind of rock.

Hollister (1974) proposes the broader term diorite model because this type of porphyry copper deposit is not restricted to island arcs nor found only in association with syenite, nor are the diorites all necessarily part of alkalic-suite plutons. Fig. 36b (page

104) gives diorite model petrography compared to other plutons having high Na_2O:-K_2O ratios. The low SiO_2:Na_2O+K_2O ratio in the host pluton significantly modifies the effect potassium and hydrogen metasomatism have on the pluton and is the basis for differences between the Lowell and Guilbert model and the diorite model.

Alteration zones within the diorite model porphyry copper deposit generally contain a core potassic assemblage with either biotite or chlorite dominant but both can occur.

Orthoclase may be entirely absent, merely less conspicuous than plagioclase (e.g., Ingerbelle or Cariboo Bell, British Columbia), or prominent, as at Stikine (Galore Creek), British Columbia. Biotite, chlorite, or both characterize the potassic zone therefore and the name potassic zone should be retained even though albite and chlorite (e.g., Afton, British Columbia) characterize the zone in some deposits. To propose a distinct zone name for occurrences where sodium silicates are dominant over potassic ignores the broad area between the extremes of sodium or potassium metasomatism. To drop the term potassic zone altogether in those diorite model deposits deficient in K-feldspar but still containing biotite would create needless confusion in the literature.

Sulfur in the hydrothermal fluids in dioritic rocks is inadequate to consume all iron in the host rocks as pyrite and iron persists throughout the hydrothermal system as chlorite and/or magnetite. Conceivably both could occur in different zones in the same system. Excess iron not consumed as pyrite or a silicate tends to persist as magnetite. The sericite expected in the phyllic zone is poorly developed or missing and is replaced by a chlorite-rich hydrothermal mineral assemblage due to incomplete removal of the iron as pyrite. The propylitic zone therefore adjoins the potassic. In this case the phyllic and argillic zones of the Lowell and Guilbert model could not develop and would not be present. The alteration zonal sequence of the diorite model commonly is potassic-propylitic because chlorite-dominant mineral assemblages of the propylitic zone commonly replace the quartz-sericite mixtures of the phyllic zone. Examples of this zoning can be seen at El Arco, Mexico, and Afton and Ingerbelle in southern British Columbia. Chlorite is the most common secondary silicate found in most deposits. However a great variety of hydrated sodium and iron-bearing silicates may be found with it. Scapolite is common in the Copper Mountain-Ingerbelle (British Columbia) area for example and other hydrated iron-bearing silicates have been reported from other deposits.

High gold:copper but low molybdenum:copper ratios are found in most diorite model deposits relative to the Lowell and Guilbert model deposits. Hypogene copper sulfide occurs in the potassic zone as dissemination and as fracture filling. Copper sulfide may also occur in the alteration zone immediately adjacent to the potassic zone and if this is a propylitic assemblage copper sulfide may occur in the propylitic zone as well as the potassic.

The paragenetic sequence of hypogene sulfide deposition in diorite model porphyry deposits shows pyrite to be early, followed by pyrite-base metal sulfide mixtures or base metal sulfide-magnetite mixtures. Substantial supergene sulfide concentrations commonly do not form in diorite model deposits. Copper tends to be fixed in place during weathering because reactive silicates neutralize the hydrogen ion generated.

Comparison of the Diorite and the Lowell and Guilbert Models

Table 10 (page 81) summaries the principal features of each model. The features noted in this table summarize the effect of hydrothermal fluids on two distinct chemical environments. The hydrothermal fluids affecting the two models could be grossly similar but the effects produced contrast markedly. Original composition of the intrusion plays an important part in the ultimate characteristics of the alteration zone. Where intrusions are silica-rich, low in calcium-magnesium, and high in sulfur potential, quartz, sericite, and pyrite may be developed in abundance. fracturing and brecciation permit development of a more intense and extensive stock-work pattern of quartz sulfide-rich veins and veinlets, and pervasive pyrite forms a typical halo around the center of mineralization. An appreciable percentage of the ore sulfides observed in the Lowell and Guilbert model porphyry copper deposits occur as vein fillings. Sulfides in the diorite model tend to occur more commonly as disseminations rather than veinlets.

Veinlets in the diorite model do not necessarily carry quartz with the sulfides; in some deposits carbonates, zeolites, epidote, K-feldspar, biotite, or other silicates are more abundant. Both the alteration products and the gangue in the veinlets of the diorite model reflect activity of hydrothermal fluids upon a diorite or quartz-deficient host. Copper minerals may occur prominently in the phyllic zone of the Lowell and Guilbert model. The phyllic zone commonly is missing in the diorite type and if this is the case copper minerals occur in both the potassic and propylitic zones. Lack of a pronounced pyrite halo in a diorite model deposit should not be interpreted as lack of metallization in the diorite porphyry. More likely it represents an inadequacy of sulfur in the hydrothermal system.

High gold, characteristci near-absence of molybdenum, and low silica content in the diorite model are compatible with porphyry copper generation in which sialic crustal contamination is minimal.

Cox (1973) and Cox, et al. (1975) describe porphyry copper deposits in Puerto Rico developed with quartz-rich quartz diorite plutons, the characteristics of which appear similar to deposits in the continental margin of Alaska. These quartz diorite-related deposits have either a chalcopyrite-magnetite or a chalcopyrite-pyrite association generally with high gold and low molybdenum values. Though proposed to be of the Lowell and Guilbert model according to definitions in this volume, these deposits may have some metallogenic affinities to the diorite model. Many other examples may be cited that combine characteristics of both models, so the models proposed here have only general application in the classification of porphyry copper deposits.

Diorite porphyry copper deposits occur in the Canadian Cordillera in a rift environment (Fox, 1975), spatially close to outcropping oceanic crust (see Chap. 5). Alaskan diorite model porphyry copper deposits are comagmatic with or penetrate Pennsylvanian-Permian marine andesitic volcanic rocks. Oceanic crust underlies these extrusives (Hollister, et al., 1975).

Lowell and Guilbert model porphyry copper deposits also may occur in close proximity to diorite type deposits. However, Lowell and Guilbert model deposits occur invariably where Precambrian cratonic crust may reasonably be inferred at depth. The silica-saturated, high molybdenum, low gold association characteristic of most deposits in this model is compatible with a mineralized intrusion contaminated with continental crustal material. Rogers, et al. (1974) show the western limit of the Precambrian craton to lie west of most of the important porphyry copper deposits of the southern Cordilleran orogen. Most deposits east of this limit may be described by the proposed definition of the Lowell and Guilbert model and most have recoverable molybdenum. Deposits in the Andean orogen typically have a prominent molybdenum content and all fall within the definition of the Lowell and Guilbert model used in this volume. The Andes porphyry copper belt is underlain by Precambrian rocks. The Appalachian orogen has been characterized as ensialic (this volume). Not surprisingly, porphyry deposits in this orogen occur within parameters of the Lowell and Guilbert model and the molybdenum:copper ratio is abnormally high.

A correlation does appear to emerge among sialic crustal thickness, metallogenic composition, and the model of porphyry at the present-day surface.

Fluid Inclusion Studies

Fluid inclusion studies have been published on Lowell and Guilbert model deposits that are useful in determining the chemistry of the hydrothermal fluids as well as the physical environment at the time of copper sulfide mineral deposition. Homogenization data from fluid inclusion studies by Nash (1976), Moore and Nash (1974), and Nash and Cunningham (1974) permit estimation of the depth at which the deposit formed and the temperature present during formation. The present surface of some deposits in the southern Cordilleran orogen in the 30-70 m.y. age range have been found to represent a depth of 1829 m (6000 ft) at time of mineral deposition.

Without exception the results of fluid inclusion studies have been interpreted to mean that the porphyry generated a highly saline brine. Temperatures have been qualitatively reported from 300-600°C in the zone that contains hypogene copper sulfide. The potassic alteration zone is included even though hypogene copper may be nearly absent in this part of the system in some deposits. Transport of metals in the highly saline brine is consistent with observations in all deposits studied. A potassic zone characterized by abundant halite-bearing inclusions is compatible with the presence of secondary silicates believed to be products of metasomatism by high-salinity fluids.

Generally smaller numbers of halite-bearing inclusions indicating moderate-salinity fluids dominate in the lower temperature quartz-sericite-pyrite alteration zone (Nash, 1976). Mixing of magmatic water with ground water may have caused the dilution as well as the observed temperature decrease. Hypogene copper content decreases consistently with the decrease in salinity of fluid inclusions.

Interpretation of Isotope Data

Isotope studies have been completed on a number of deposits in the Western Hemisphere and those pertinent to deposits of the southern Cordilleran orogen have been described in that chapter. These provide insight into the genesis of this type of deposit. They may be briefly summarized for each element that has been investigated.

Sulfur Isotope Studies: Cheney (1974) and Rye and Ohmoto (1974) summarize the isotope data now available for sulfur from sulfide in what this volume defines as the Lowell and Guilbert model porphyry copper deposit. Both acknowledge that sulfide sulfur in most porphyry copper systems is derived most importantly from deep-seated sources. The S^{34}/S^{32} ratios of these deposits suggest an upper mantle or homogenized crustal source for the sulfur. Sulfur is intimately associated with porphyry copper type deposits as a dissemination in the pluton and the element was probably derived from the host igneous rock. At least part of the rock itself in the Lowell and Guilbert model therefore is indicated to be derived from the upper mantle or homogenized crust. Should the speculation be accepted that copper also is largely derived from the same source as sulfur, then an upper mantle or homogenized crustal source also is suggested for some of this metal.

A paucity of published isotope data on sulfide sulfur for diorite model deposits does not permit a projection as to origin of sulfur in that model. If diorite model deposits tend to form coincident with a thin continental crust it would be surprising if sulfide sulfur data for diorite model deposits differed significantly from that found in Lowell and Guilbert model deposits.

Lead Isotope Studies: Zartman (1974) and Doe and Stacey (1974) have summarized data from lead isotope studies that bear on the origin and genesis of Lowell and Guilbert model porphyry copper deposits as defined here. The lead isotopic composition of copper sulfide is similar to that of the associated igneous rocks.

The more enriched in radiogenic lead isotopes a particular deposit is, the greater has been the involvement and contamination with upper crustal sources; the least radiogenic leads in rocks are assumed to contain the least component from the upper crust. Lowell and Guilbert model porphyry copper deposits as a whole have the least amounts of radiogenic leads of any hydrothermal ore deposits although the plutons have lead isotopes that reflect contamination from the sialic crust penetrated. The conclusion that lead in the pluton and in the hypogene copper sulfides came from the same source and that upper crustal material may have contributed to both seems justified from lead studies. No lead isotopic studies are yet published on diorite model deposits.

Oxygen and Hydrogen Isotope Studies: Taylor (1974) and Sheppard, et al. (1971) show that the D/H (deuterium/hydogen) ratio and the O^{18} shift may be used to define sources of hydrothermal fluids and to qualitatively estimate the temperature of mineral deposition in porphyry copper deposits. At present studies have been published only on deposits of the Lowell and Guilbert model. The D/H ratio and the O^{18} values found in hydrothermal biotites from the potassic zone of porphyry copper deposits form a very tight grouping with normal magmatic biotites. Oxygen and hydrogen isotope studies on hydrothermal biotites suggest potassic zone mineralization to be derived from the compositional field of magmatic water. Qualitatively, temperatures have been reported from 450-600°C in this zone based on oxygen isotope thermometry.

D/H ratios and the O^{18} shift for sericite and clay minerals of the phyllic zone on the other hand have values that require the presence of a meteoric- or saline-formation water component. Temperatures from 300-450°C for the phyllic zone have qualitatively been reported.

D/H ratios and the O^{18} shift for argillic and propylitic zone minerals show even less involvement of the magmatic-hydrothermal fluid in development of these assemblages. Oxygen and hydrogen isotopic data therefore clearly support the contention that a central magmatic-hydrothermal system gives way to an external meteoric-hydrothermal system. As distance is gained from the center of mineralization involvement by constituents of the magmatic-hydrothermal system sharply decreases and the temperature drops.

Magmatic-hydrothermal fluids therefore seem likely to have formed during the late stages of crystallization in the upper portions of a porphyry stock to give rise to the potassic zone. Outside this zone a meteoric-hydrothermal circulation is established that forms the phyllic, argillic, and propylitic zones. Both systems are simultaneously present in the early history of the deposit but the external system will persist after the magmatic-hydrothermal system fades. Persistence of the external system after the internal system ceases may lead to the superposition of products of the external system over the internal.

Although deposits of the diorite model have not yet been studied for D/H ratios and O^{18} shift, the data for deposits of this model are anticipated to remain the same.

Strontium Isotope Studies: Moorbath, et al. (1967), Lowell (1974), Hedge (1974), and Kesler, et al. (1975) have all presented data on initial Sr^{87}/Sr^{86} ratios in the Lowell and Guilbert model porphyry copper plutonic systems. These data show the ratios thus far determined to be compatible with derivation of the magma from the mantle with some contamination evident from the crust. Moorbath, et al. (1967) stress a mantle source, whereas Hedge (1974) points out the evidence of contamination.

Kesler, et al. (1975, p. 524) present a summary of initial Sr^{87}/Sr^{86} ratios for intrusive rocks associated with Lowell and Guilbert model (as defined in this volume) porphyry copper deposits. Their study of 16 deposits shows a significant break occurring at an initial ratio of about 0.7055. The seven deposits in their study with ratios greater than 0.7055 are all from a cratonic environment and have significant byproduct molybdenum with the copper. The other nine deposits with initial ratios of less than 0.7055 are largely from a continental margin or an island arc environment where cratonic conditions do not exist. Although exceptions appear to occur the limited data available at present suggest molybdenum is generally not important in deposits with an initial Sr^{87}/Sr^{86} ratio of less than 0.7055.

Halpern (1975) provides initial Sr^{87}/Sr^{86} ratios from a total of nine rock samples of copper-bearing porphyry from the El Salvador (Chile) deposit that range between 0.7025 and 0.7037 with a mean of 0.7030. Halpern interprets the El Salvador values to support the hypothesis that copper-rich porphyries are related genetically to subducted oceanic lithosphere.

Ratios for diorite model deposits have not yet been published but it is anticipated that initial Sr^{87}/Sr^{86} ratios for them would be less than 0.7055.

Rb/Sr ratios also appear to differ among Lowell and Guilbert model deposits. Those from cratonic settings average 0.3 to 0.4 whereas others from a noncratonic continental margin or island arc average about 0.2. Diorite model deposits are forecast to have 0.2 Rb/Sr ratios.

Conclusions Derived from Isotope Data: The sulfur, rubidium-strontium, and lead data all point to what appears to be a contaminated upper mantle source for at least those elements in the stock present in most porphyry copper deposits. It seems simple to relate much of the stock itself to the same source. The lead isotope ratios in hypogene copper minerals indicate the metal had the same source as the pluton. Distribution of copper in the potassic and phyllic zones further suggests this metal originated with the magma and rose into the upper crust during ascent of the intrusion. Most Lowell and Guilbert model porphyry copper deposits that are richest in molybdenum are confined to those areas of continental crust that, if not cratonic in character, were thick and included at least Paleozoic metamorphics.

The diorite model porphyry copper deposits having the most notable gold content (but the least molybdenum) penetrated thick sections of marine volcanics that may have included oceanic crust but the presence of a substantial continental crust as a host for this model cannot be demonstrated.

The conclusion suggested by these associations is that at least part of the minor metals associated with the porphyry copper deposits (such as molybdenum and gold) may have been acquired as contaminants by the magma and its fluids as they ascended through the crust.

Greenwood (1975) offers a model for magma generation that appears compatible with physical and isotopic evidence found in porphyry copper deposits. According to his concept magma is initially derived from partial melting of a solid parent rock. Peridotite and lherzolite in the upper mantle may supply the initial melt if less than 10% and more probably 5% of the rock is fused in the presence of, carbon dioxide or water. Dry partial melting of a gabbroic eclogite provides a similar initial melt. Rare earth and alkali fractionation curves from inclusions of these rocks in basalt indicate that about 5% fusion is required to explain these ultramafic rocks as residues.

Partial fusion of the peridotite and lherzolite may take place in a low velocity zone in the upper mantle. The 5% liquid by weight present in this zone could be free to move only in the presence of deformation extending to the mantle. A Benioff zone could provide such a deformational environment but it is conceivable that any fracture permitting the liquid phase of the low velocity zone to mobilize and escape would be a satisfactory vehicle for germination of a porphyry copper deposit. Since the amount of liquid produced is about only 5% by weight the extractive process must be highly efficient. The resulting upward concentration of those fractionated rock-forming elements (potassium, sodium, rubidium, strontium) would be accompanied by lead, uranium, copper, and sulfur but not cobalt and nickel. Strong fractionation of these elements into the melt and emplacement of this magma into the lower crust permit the magma to act as a heat-transfer mechanism, fusing parts of preexisting rocks.

Andesitic magmas could be produced by the reaction of the ascending melt and a gabbroic eclogite. Andesite is an extrusive commonly found in volcanic piles that form part of a porphyry copper-bearing complex (e.g., Highland Valley, BC; Mariner, NS; Silver Bell, AZ, and numerous Andean examples). The diorite model association also could be derived from such an evolution.

Andesitic magmas produced could ascend and become emplaced in the crust. These magmas could assimilate part of the crust and also would differentiate. Emplacement with melting and differentiation provide the variety of igneous rocks found in some volcanic centers that also contain porphyry copper deposits. The introduction of a substantial part of the molybdenum into the system took place at this time.

Such an evolutionary path would permit the formation of the variety of intrusive rocks known to accompany some porphyry copper deposits. The crust in the island arc and continental margin areas could be expected to provide magmas with high Na_2O: K_2O ratios whereas a crust that included a Precambrian craton or a thick sialic section would produce plutons richer in potassium and quartz. The Lowell and Guilbert model as proposed in this volume approximates or represents deposits found with a quartz-bearing magma regardless of the type of crust involved. The diorite model deposit is associated with a quartz-deficient magma that appears to lack most of the silica and molybdenum contamination ascribed to the sialic crust.

Large batch melting of source rock (Wyllie, et al., 1976), distinct from the partial melting mechanism, provides one explanation for the evolution of batholiths. This type of melting, however, gives little relative concentration or relative enrichment of alkalies and trace elements and is not effective in concentrating copper.

Isotope Inferences for the Lowell and Guilbert Model, Stockwork Type: Fig. 56 shows the Lowell and Guilbert model as a stockwork type deposit associated with a pluton that did not reach the surface. In this general example the area of fractured rock is shown in and around the pluton and the alteration zones tend to be confined to its vicinity. The fractures may be genetically related to the intrusion although this is not clearly shown in Fig. 56. Fractures are confined to the area of the pluton however and provide access for fluids driven by the pluton as a heat source.

As suggested by Taylor (1974), highly saline magmatic-hydrothermal fluids formed the potassic zone during late stage crystalli-

zation of the stock. At Ray and a few other deposits presence of the potassic zone outside the stock infers that magmatic-hydrothermal fluids can exist well outside the stock as well as in its interior.

Meteoric-hydrothermal circulation becomes established outside the potassic zone. The action of these thermal fluids results in concentric alteration zoning typical of the Lowell and Guilbert (1970) model—i.e., phyllic, argillic, and propylitic. Breccia pipes occurring in these alteration zones are also smaller structures and are found where stockwork fractures, which provide pressure relief for fluids, are poorly developed.

Both the magmatic-hydrothermal and the meteoric-hydrothermal systems exist together within the porphyry copper during its early stages. As the potassic zone is completely developed and the activity of magmatic-hydrothermal fluids fades the meteoric-hydrothermal fluids will tend to invade the potassic zone as this convection cycle persists while the heat source cools. Phyllic zone alteration therefore may be superimposed on the potassic zone. Those deposits lacking a potassic zone may have developed such a zone early; a phyllic zone may have collapsed on it, masking or destroying the potassic zone.

Fluid inclusion studies that infer base metal sulfide deposition to originate from a highly saline brine coupled with the early paragenetic appearance of pyrite are compatible with the notion that base metals are introduced into a pyrite-bearing rock as halides. Base metal sulfides could be deposited in tectonic openings as a result of reaction between the highly saline brine and the pyrite-bearing wall rock. The copper and some of the other metals would be transported by the brine into the zone with pyrite, with precipitation of the metals as sulfides. The late introduction of metals into the system is consistent with field evidence. Consistent association between veins and alteration halos in the vein walls (Sales and Meyer, 1948) is compatible with this hypothesis.

If the highly saline magmatic-hydrothermal fluid provides the vehicle whereby meta-

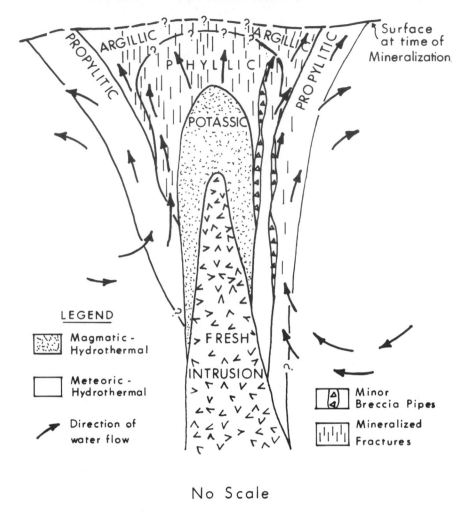

Fig. 56. Lowell and Guilbert model, stockwork type. The hypothetical development of alteration zoning in a stockwork type deposit of the Lowell and Guilbert (1970) model is shown in this figure for a mineralized intrusion that did not reach the surface. In this example the meteoric-hydrothermal system did not collapse on and destroy the potassic zone developed by magmatic-hydrothermal fluids. Arrows in this diagram indicate fluid flow during the coincident development of both potassic and phyllic zones.

somatic changes occur, developing potassic zone silicate and sulfide mineralogy, then as the magmatic-hydrothermal fluid escapes the core it conceivably could carry sulfide ion with it as a result of reactions in the potassic zone. Outward flow of the magmatic-hydrothermal fluid and its mixing with a meteoric-hydrothermal fluid in a cooler environment could lead to the concentration of pyrite observed in the phyllic zone where the chemical and physical environment encourages such deposition. In this hypothesis iron and sulfur leached from the potassic zone would crystallize in the phyllic. Mineralogy of the phyllic zone is consistent with conditions that favor pyrite concentration. Fluid inclusion studies commonly show a more moderate brine content during mineralization in this zone. Pyrite also is concentrated in most deposits in the outer part of the phyllic zone, outside the zone of strongest hypogene copper. Leaching sulfide ion from the potassic zone to concentrate pyrite outside this zone is also compatible with the formation of magnetite in the potassic zone.

Mineralization is zoned outward from a copper-molybdenum core with sphalerite, galena, base metal sulfosalts, and precious metals zonally arranged about the heat source. Alteration minerals present in the outer deposits suggest that the sulfides are deposited in fractures as a result of activity

of meteoric-hydrothermal dominant systems.

Isotope Inferences for Lowell and Guilbert Model, Major Breccia Pipe Type: Fig. 57 depicts the general relationship in major breccia pipes such as Toquepala (Richard and Courtright, 1958), Braden (Howell and Molloy, 1960), and La Caridad (Saegart, et al., 1974). Where tourmaline is present the distinction between the fragments and the cementing material is emphasized. If a

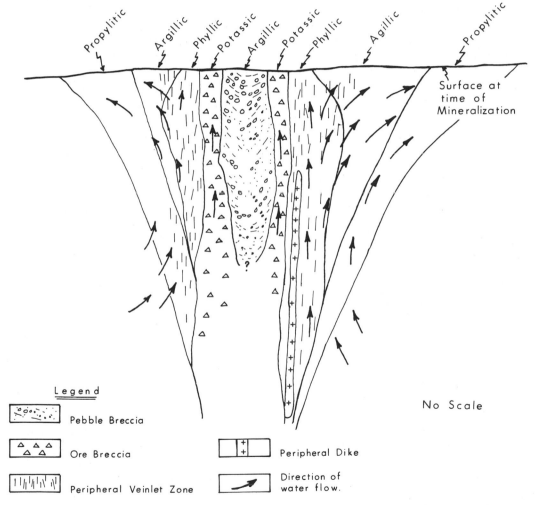

Fig. 57. Lowell and Guilbert model, breccia type. In a major breccia pipe or breccia type porphyry copper deposit the magmatic-hydrothermal system may precede and then be active coincident with a meteoric-hydrothermal system generated by the heat source. If the meteoric-hydrothermal system supercedes the magmatic-hydrothermal, material from the potassic and other zones may be incorporated in a late low-temperature phase such as a pebble breccia. Arrows in this diagram indicate fluid flow at the time of coincident development of potassic and phyllic zones.

postsulfide pebble breccia is present, the presence of cobbles of an earlier ore breccia embedded (commonly with angular fragments) is easily discernable as rebrecciated and rounded fragments in the later pebble pipe or rounded fragment breccia. The later pebble breccia is argillic dominant and has crude graded bedding visible. Ore sulfides form part of the cement of the ore breccia while pebble breccia cement is largely postore clay.

Potassic zone silicates, usually characterized by biotite and orthoclase, with ore sulfides commonly are the dominant alteration products in fragments of ore breccias. They may also exist as cementing material for the fragments. Isotopic studies on stockwork porphyry copper potassic zones suggest that magmatic-hydrothermal fluids are responsible for this alteration suite. These fluids ascended under strong lithostatic pressure during and immediately subsequent to formation of the pipe.

The presence of boron in this ascending magmatic-hydrothermal fluid permits the path of the solution to be traced as it begins to mingle with meteoric water in a convective circulation system. Thus the development of tourmaline in the propylitic, argillic, and phyllic alteration zones outside the pipe where isotope studies indicate meteoric-hydrothermal fluid to have been important suggests that some components of the magmatic-hydrothermal fluids were present in the meteoric-hydrothermal system. In the large breccia pipe the heat source conceivably was largely that heat carried with the magmatic-hydrothermal fluids, since large epizonal stocks are not present for each pipe at or near the surface.

If evidence cited elsewhere in this volume is accepted the paragenetic sequence of events in breccia pipe mineralization would appear to be:

1) Early introduction of tourmaline-quartz-pyrite mineralization.

2) Development of base metal mineralization with fluid inclusion studies showing hypogene copper sulfide to be associated with a highly saline fluid.

This sequence of events is compatible with the hypothesis that base metal sulfides were deposited in voids between breccia fragments as the magmatic-hydrothermal fluid reacted with these rocks to metasomatize them, precipitating base metal sulfide as pyrite in the fragments became vulnerable.

As in the stockwork deposit both the magmatic-hydrothermal and the meteoric-hydrothermal fluids were simultaneously present during one stage of the mineralizing process and in the breccia pipe initial development of the magmatic-hydrothermal system, forming the potassic alteration zone, was clearly an early event. Alteration minerals associated with the meteoric-hydrothermal solutions (propylitic, argillic, and phyllic zones) continue to develop so long as a central heat source exists, long after the magmatic-hydrothermal fluid has ceased to be an effective force. After sulfide deposition has ceased the convective circulation system continues to function in many large pipes and the pebble breccia forms in this postsulfide mineral stage (see Fig. 52). Development of graded bedding and rounded fragments suggests that convection in the meteoric-hydrothermal system, which collapsed on the pipe as it cooled, involved fluids with temperatures below boiling. As the magmatic-hydrothermal phase faded the meteoric-hydrothermal pervaded the entire structure, dying as the heat source cooled.

In this type of deposit, the sulfur, copper, boron, and lead are largely from a deep-seated source. Water has at least two major sources, magmatic (deep-seated) and meteoric.

Isotope Inferences for the Diorite Model: Fig. 58 shows the general relationships present for the diorite model porphyry copper as well as those relationships that may appear for dry examples (e.g., Ajo) of the Lowell and Guilbert model. In this generalized example, Fig. 58 shows a simple intrusion that did not reach the surface.

The potassic alteration zone, on the basis of alteration and isotope studies in Lowell and Guilbert model deposits, is typified by K-feldspar and/or biotite and/or chlorite,

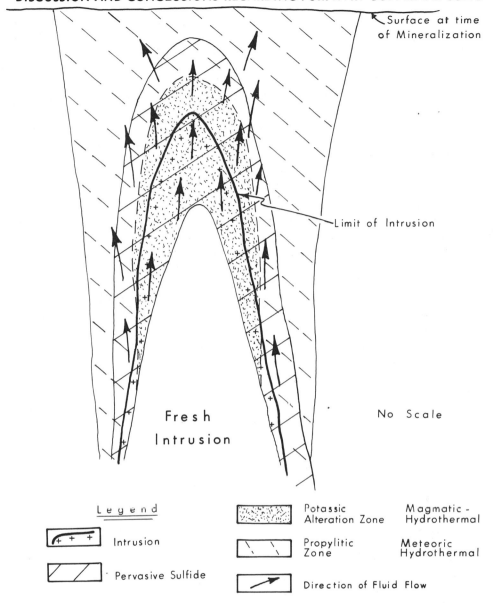

Fig. 58. Diorite model. Flow direction during simultaneous action by the magmatic-hydro-thermal and meteoric-hydrothermal systems is indicated by arrows. Activity by the latter after cessation of magmatic-hydrothermal fluid flow may result in substantial destruction of the previously formed potassic zone (as at Copper Mountain-Ingerbelle, British Columbia).

whereas the fluids responsible were domi-nantly magmatic-hydrothermal. The mag-matic-hydrothermal fluids were under strong hydrostatic pressure and invaded the host rock through fractures genetically associated with intrusion of the pluton.

Outside the stock the magmatic-hydro-thermal fluids mingled with meteoric water as convective circulation developed around the heat source. The meteoric-hydrothermal fluids developed carried some sulfur, copper, and possibly gold from the developing potas-sic zone into the surrounding host rocks. If the sulfur content of the meteoric-hydrothermal solution was inadequate to pyritize the host rock, propylitic zone minerals formed adjacent to the potassic zone.

Gold may have been mobilized by the meteoric-hydrothermal fluids from the intruded rock as well as from the intrusion but this is difficult to assess quantitatively. Fox (1975) cites examples of bornite and abnormally high gold in the potassic zone of some deposits with a chalcopyrite-pyrite gold-deficient zone surrounding the bornite zone and existing in a chlorite-epidote-rich propylitic zone peripheral to the potassic zone.

In this example water is both magmatic and meteoric in the propylitic zone although saline water from the intruded formations or seawater may be an added component. In the potassic zone magmatic water could be expected to dominate. The sulfur and copper are believed to be largely of deep-seated origin.

Plate Tectonic Considerations

Mitchell (1973), Sillitoe (1972, 1973), and others have suggested a genetic tie between subduction and formation of porphyry copper deposits. Lowell (1974) and Noble (1976) point out the difficulty in deriving southwestern US porphyry copper deposits from a subduction zone. The province has neither the shape nor orientation of a subduction zone-related phenomena. Noble (1974) and Lowell (1974) show the linear distribution of porphyry copper deposits in the southwestern US to represent some deep-seated control that is probably not subduction related. Noble (1974, p. 16) also notes that if these ore deposits were related to plate movement a progressive decrease in age eastward would be expected in the distribution of the porphyry deposits. No such relationship exists, however.

The lead isotopic characteristics of oceanic crust almost certainly rule that out as a source for both porphyry magmas and metal in a converging plate environment (Zartman, 1974).

Younker (1975) concludes that the magma flux for melting above subduction zones is a function of the rate at which heat is supplied for the fusion process and the rate at which the magma collects into bubbles of sufficient size to rise independently of the solid residue. In subduction zones both these rates are a function of plate velocity, thereby establishing a relationship between plate dynamics and the development of batholiths. Higher rates of plate movement also lead to higher shear strain and more efficient magma collection with development of batholiths.

Larson and Pitman (1972) have proposed that oceanic magnetic anomaly reversals be used to determine periods of simultaneous rapid sea floor spreading and more rapid subduction. In this hypothesis granodiorite batholith intrusion and major fold and thrust events were more abundant during periods of rapid subduction. Fig. 59 shows the magnetic anomaly reversals, the time distribution for Sierra Nevada type batholiths, and dates on porphyry copper deposits in the southern Cordilleran orogen. Very few dated porphyry deposits coincide with times suggested for rapid sea floor spreading and subduction. Most prefer to avoid the compressive environment inferred by rapid subduction.

Fig. 60 gives the dates for porphyry copper deposits and batholiths for the Andean orogen on the same magnetic anomaly reversal scale. Again, independence between the development of porphyry deposits and batholith intrusion is clearly indicated. Fig. 61 shows the same relationship for the northern Cordilleran orogen. Again, a poor correlation is found to exist between dates for porphyry copper deposits and dates for Sierra Nevada type batholiths. The inference seems clear that based on present dating the conditions that favored rapid subduction, batholith development, and large-scale thrusting and folding did not favor formation of porphyry copper deposits. The deposits tended to form during more relaxed regimes of strike-slip tectonics.

Although porphyry copper deposits may not be uniquely identified genetically with subduction, their position within the western edge of the Cordillera of North and South America does suggest that they formed as a response to plate tectonics.

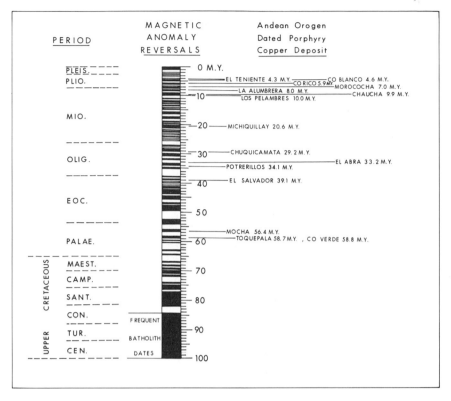

Fig. 59. Magnetic anomaly, Andes. Ages of porphyry copper deposits plotted with oceanic magnetic quiet periods and periods of maximum batholith activity in the Andes show a poor correlation between porphyry copper deposition and the other two phenomena (*modified from Larson and Pitman, 1972*).

The increased probability for all types of deep fracturing near a plate boundary improves the possibility that a porphyry copper deposit will germinate in that area. Water from material taken down a Benioff zone will enhance the possibility of magma development while the complications that arise when a low velocity zone (a zone of partial melting) is intersected by a Benioff zone should be conducive to magma generation in the mantle.

If the crust may be ascribed as a source for some of the gold and molybdenum, reliance on the subduction zone for metals is weakened and this also weakens the genetic tie between subduction and porphyry copper systems.

The conclusion that a porphyry will be germinated wherever the structural conditions will permit extraction of the seed material from the low velocity zone seems logical and these parameters do not exclude magma generation from a subduction zone.

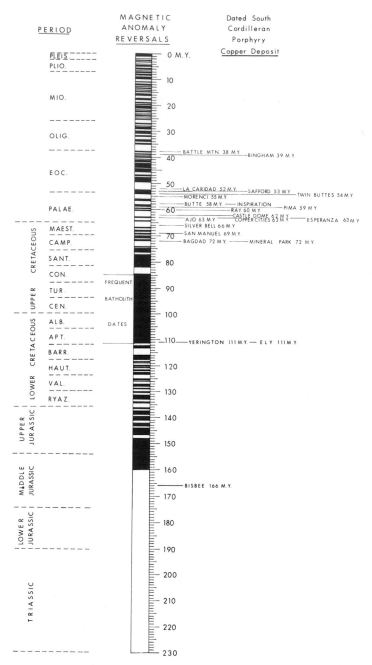

Fig. 60. Magnetic anomaly, southern Cordilleran orogen. Ages of porphyry copper deposits plotted with oceanic magnetic quiet periods and periods of maximum batholith activity in the southern Cordilleran orogen show a poor correlation between porphyry copper occurrences and the other two phenomena for the area south of the Columbia Plateau (*modified from Larson and Pitman, 1972*).

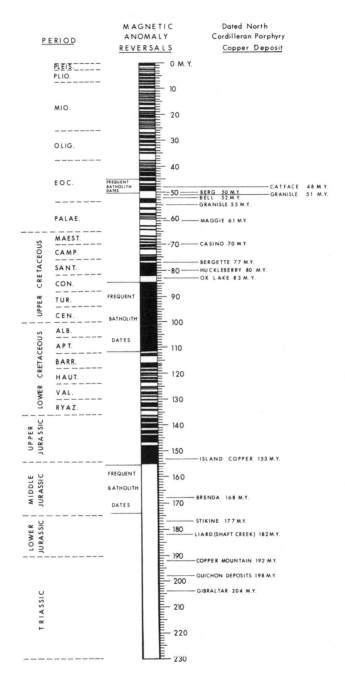

Fig. 61. Magnetic anomalies, northern Cordilleran orogen. Ages of porphyry copper development compared with magnetic quiet periods of oceanic basins and periods of batholith activity in the northern Cordilleran orogen show a poor correlation between porphyry copper development and the other two phenomena for the area from the Columbia Plateau to the St. Elias Range, Alaska (*modified from Larson and Pitman, 1972*).

REFERENCES AND BIBLIOGRAPHY

Bryner, L., 1961, "Breccia and Pebble Columns Associated with Epigenetic Ore Deposits," *Economic Geology,* Vol. 56, pp. 1-12.

Cheney, E. S., 1974, "Examples of the Application of Sulfur Isotopes to Economic Geology," *Trans., SME-AIME,* Vol. 256, pp. 31-38.

Cox, D. P., 1973, "Hydrothermal Alteration in Puerto Rican Porphyry Copper Deposits," *Economic Geology,* Vol. 68, p. 1329.

Cox, D. P., Gonzales, I. P., and Nash, J. T., 1975, "Geology of the Sapo Alegre Porphyry Copper Prospect," *Journal of Research,* US Geological Survey, Vol. 3, p. 313.

Doe, B. R., and Stacey, J. S., 1974, "The Application of Lead Isotopes to the Problems of Ore Genesis and Ore Prospect Evaluation," *Economic Geology,* Vol. 69, pp. 757-776.

Fox, P. E., 1975, "Alkaline Rocks and Related Mineral Deposits of the Quesnel Trough, British Columbia," *Abstracts,* Symposium on Intrusive Rocks and Related Mineralization of the Canadian Cordillera, Geological Association of Canada, Cordilleran Sec.

Greenwood, H. J., 1975, "Genesis of Magmas," *Abstracts,* Symposium on Intrusive Rocks and Related Mineralization of the Canadian Cordillera, Geological Association of Canada, Cordilleran Sec.

Halpern, M., 1975, "Strontium Isotope Composition of Magmatic Rocks from Northern Chile," *Abstracts,* Geological Society of America, Vol. 7, p. 1096.

Hedge, C. E., 1974, "Strontium Isotopes in Economic Geology," *Economic Geology,* Vol. 68, pp. 823-826.

Hollister, V. F., 1974, "Regional Characteristics of Porphyry Copper Deposits of South America," *Trans., SME-AIME,* Vol. 256, pp. 45-53.

Hollister, V. F., 1976, "Evolution of the Porphyry Copper Province of the Northern Cordilleran Orogen," *Proceedings,* 1st Circum-Pacific Mineral Resources Conference, American Association of Petroleum Geologists.

Hollister, V. F., Anzalone, S. A., and Richter, D. H., 1975, "Porphyry Copper Belts of Southern Alaska and Adjacent Yukon," Bulletin No. 4, Canadian Institute of Mining and Metallurgy, pp. 63-70.

Howell, F. H., and Molloy, J. S., 1960, "Geology of the Braden Ore Body, Chile, South America," *Economic Geology,* Vol. 55, pp. 863-906.

Joralemon, I. B., 1952, "Age Cannot Wither *or* Varieties of Geologic Experience, *Economic Geology,* Vol. 47, p. 253.

Kents, P., 1964, "Special Breccias Associated with Hydrothermal Developments in the Andes," *Economic Geology,* Vol. 59, p. 1551.

Kesler, S. E., Jones, L. M., and Walker, R. L., 1975, "Intrusive Rocks Associated with Porphyry Copper Mineralization in Island Arcs," *Economic Geology,* Vol. 70, p. 515.

Kuhn, T. H., 1941, "Pipe Deposits of the Copper Creek Area, Arizona," *Economic Geology,* Vol. 36, p. 512.

Larson, R. L., and Pitman, W. C., 1972, "World Wide Correlation of Mesozoic Magnetic Anomalies," *Bulletin,* Geological Society of America, Vol. 83, p. 3645.

Locke, A., 1926, "The Formation of Certain Ore Bodies by Mineralization Stoping," *Economic Geology,* Vol. 21, pp. 421-453.

Lowell, J. D., and Guilbert, J., 1970, "Lateral and Vertical Alteration Mineralization Zoning in Porphyry Ore Deposits," *Economic Geology,* Vol. 65, No. 4.

Lowell, J. D., 1974, "Regional Characteristics of Porphyry Copper Deposits of the Southwest," *Economic Geology,* Vol. 69, pp. 601-617.

Mills, J. W., 1972, "Origin of Copper-Bearing Breccia Pipes," *Economic Geology,* Vol. 67, p. 533.

Mitcham, T. W., 1974, "Origin of Breccia Pipes," *Economic Geology,* Vol. 69, p. 412.

Mitchell, A. H. G., 1973, "Metallogenic Belts and Angle of Dip of Benioff Zones," *Nature (London), Physical Science,* Vol. 24t, No. 143, pp. 49-52.

Moorbath, S., Hurley, P. M., and Fairbairn, H. W., 1967, "Evidence for the Origin and Age of Some Mineralized Laramide Intrusives in the Southwestern United States from Strontium Isotope and Rubidium-Strontium Measurements," *Economic Geology,* Vol. 62, pp. 228-236.

Moore, W. J., and Nash, J. T., 1974, "Fluid Inclusion Studies at Bingham Canyon," *Economic Geology,* Vol. 69, No. 5, p. 631.

Nash, J. T., and Cunningham, C. G., 1974, "Fluid Inclusion Studies of the Porphyry Copper Deposits at Bagdad, Arizona," *Journal of Research,* US Geological Survey, Vol. 2, p. 31.

Nash, J. T., 1976, "Fluid Inclusions as a Guide to Porphyry Copper Deposits," Open-File Report No. 76-482, US Geological Survey.

Ney, C. S., and Hollister, V. F., 1976, "Geologic Setting of the Porphyry Deposits of the Canadian Cordillera," *Porphyry Deposits of the Canadian Cordillera,* Special Vol. 15, Canadian Institute of Mining and Metallurgy.

Noble, J. A., 1974, "Metal Provinces and Metal Finding in the United States," *Mineralium Deposita,* Vol. 9, pp. 1-25.

Noble, J. A., 1976, "Metallogenic Provinces of the Cordillera of Western North and South America," *Mineralium Deposita,* Vol. 11, pp. 219-233.

Perry, V. D., 1961, "The Significance of Mineralized Breccia Pipes," *Mining Engineering,* No. 10, p. 367.

Portacio, J. S., 1974, "Notes on Hydrothermal Alteration in Philippine Porphyry Copper Deposits," 4th Symposium on Mineral Resource Development, Sec. 1, Dec. 6-7, Manila, Philippines.

Richard, K., and Courtright, J. H., 1958, "Geology of Toquepala, Peru," *Mining Engineering,* No. 10, pp. 262-266.

Rogers, J. J. W., et al., 1974, "Paleozoic and Lower Mesozoic Volcanism and Continental Growth in the Western United States," *Bulletin,* Geological Society of America, Vol. 85, pp. 1913-1920.

Rye, R. O., and Ohmoto, H., 1974, "Sulfur and Carbon Isotopes and Ore Genesis: A Review," *Economic Geology,* Vol. 69, pp. 826-842.

Saegart, W. E., Sell, J. D., and Kilpartick, B. E., 1974, "Geology and Mineralization of La Caridad Porphyry Copper Deposit, Sonora, Mexico," *Economic Geology,* Vol. 69, pp. 1060-1077.

Sales, R. H., and Meyer, C., 1948, "Wall Rock Alteration at Butte, Montana," *Trans., AIME,* Vol. 178, pp. 9-35.

Sheppard, S.M.F., Nielsen, R. L., and Taylor, H. P., 1969, "Oxygen and Hydrogen Isotope Ratios of Clay Minerals from Porphyry Copper Deposits," *Economic Geology,* Vol. 64, pp. 755-777.

Sillitoe, R. H., and Sawkins, F. J., 1971, "Geologic, Mineralogic and Fluid Inclusion Studies of Tourmaline Breccia Pipes, Chile," *Economic Geology,* Vol. 66, pp. 1028-1041.

Sillitoe, R. H., 1972, "A Plate Tectonic Model for the Origin of Porphyry Copper Deposits," *Economic Geology,* Vol. 67, pp. 184-197.

Sillitoe, R. H., 1973, "Geology of the Los Pelambres Porphyry Copper Deposit, Chile," *Economic Geology,* Vol. 68, p. 1.

Soregaroli, A. E., 1975, "The Geology of Molybdenum and Copper Deposits in Canada," *Canada,* Paper No. 75-1, Pt. A, Geological Survey of Vancouver.

Sutherland Brown, A., 1969, "Mineralization in British Columbia and the Copper and Molybdenum Deposits," *Transactions,* Canadian Institute of Mining and Metallurgy, Vol. 72, pp. 1-15.

Sutherland Brown, A., 1974, "Metallogeny in the Canadian Cordillera," Pacific Rim Resources Conference, Honolulu, HI, Aug.

Taylor, H. P., 1974, "Application of Oxygen and Hydrogen Isotope Studies," *Economic Geology,* Vol. 69, pp. 843-883.

Titley, S. R., 1972, "Intrusion and Wall Rock Porphyry Copper Deposits," *Economic Geology,* Vol. 67, p. 122.

Wyllie, P. J., et al., 1976, "Granitic Magmas," *Canadian Journal of Earth Sciences,* Vol. 13, pp. 1007-1019.

Younker, L. W., 1975, "Relationship Between Plate Dynamics and Development of Batholiths," *Abstracts,* Geological Society of America, Vol. 7, p. 1326.

Zartman, R. E., 1974, "Lead Isotopic Provinces in the Cordillera of the Western United States and Their Geologic Significance," *Economic Geology,* Vol. 69, pp. 792-805.

Index